Synthesis Lectures on Information Concepts, Retrieval, and Services

Series Editor

Gary Marchionini, School of Information and Library Science, The University of North Carolina at Chapel Hill, Chapel Hill, USA

This series publishes short books on topics pertaining to information science and applications of technology to information discovery, production, distribution, and management. Potential topics include: data models, indexing theory and algorithms, classification, information architecture, information economics, privacy and identity, scholarly communication, bibliometrics and webometrics, personal information management, human information behavior, digital libraries, archives and preservation, cultural informatics, information retrieval evaluation, data fusion, relevance feedback, recommendation systems, question answering, natural language processing for retrieval, text summarization, multimedia retrieval, multilingual retrieval, and exploratory search.

Rong Tang · Graham Herrli

The Domain of UX
in Information Studies:
Bridging Theories, Research,
and Professional Practice

 Springer

Rong Tang
Simmons University
Boston, MA, USA

Graham Herrli
Mountain View, CA, USA

ISSN 1947-945X   ISSN 1947-9468 (electronic)
Synthesis Lectures on Information Concepts, Retrieval, and Services
ISBN 978-3-031-83528-5   ISBN 978-3-031-83529-2 (eBook)
https://doi.org/10.1007/978-3-031-83529-2

This Springer imprint is published by the registered company Springer Nature Switzerland AG
The registered company address is: Gewerbestrasse 11, 6330 Cham, Switzerland

If disposing of this product, please recycle the paper.

To my father, Yongcheng Tang, who passed away in September 2023. His unwavering encouragement and belief inspired me to strive for excellence and approach my work with diligence and purpose. To my husband, Heyun Yin, whose steadfast support made the book's completion possible.

—Rong Tang

Contents

List of Figures

List of Tables

Introduction

1

Overview

This book was inspired by the experiences of both authors in their years of involvement in academic research and teaching, as well as conducting User Experience (UX henceforth) research and design in work practices. As a result of their experience in the field, both authors observe differences in the expectations, accepted norms, and outcomes of UX research between the academic side and the applied side. The degree of separation of applied and academic UX research is alarming and has not been adequately addressed by the UX community. In two international studies surveying UX researchers in academic and practical settings, Law et al. (2009) and Lallemand et al. (2015) found that a wide gap exists in definitions, understanding of the nature of UX work, and methodologies between the two domains. This disconnect includes a lack of synergy between research discoveries in the academic realm and research discoveries in applied settings. It also results in localized and segmented UX efforts, with findings shared only within a circle of UX researchers. Because both scholarly and practical UX efforts tend to stay within a particular setting, much work has been done separately with duplicated efforts and missed opportunities for cross-pollination of knowledge and advancements across the two domains.

In his review of 100 years of the UX field, Jakob Nielsen (2017) acknowledged that while the century-long perspective is "positive beyond belief" and "UX has come a long way since 1950," he still feels "the overall quality of user experience… is less than 10% of what it should be."

For UX research to truly advance, and for the UX communities to thrive, it is crucial to identify points of convergence between academic and applied UX research. As Lallemand et al. (2015) argue, "the numerous differences observed between Industry and Academia

© The Author(s), under exclusive license to Springer Nature Switzerland AG 2025   1
R. Tang and G. Herrli, *The Domain of UX in Information Studies: Bridging Theories, Research, and Professional Practice*, Synthesis Lectures on Information Concepts, Retrieval, and Services, https://doi.org/10.1007/978-3-031-83529-2_1

clearly indicate that there is however still a gap between both perspectives. A better integration of theories and practice should thus be a primary goal, undoubtedly leading to a win–win situation for both Academia and Industry" (p. 47). In this book, after discussing the gaps in UX research in terms of definitions, models, methods, and reporting mechanisms, in Chapter "Where to Go from Here: Bridging the Gaps", we propose a framework outlining actions to bridge these divides.

Stories Indicating the Need to Bridge Between Academic and Applied UX

Both authors have unique perspectives and experiences regarding the divide between academic and applied UX, and thus have different stories to share.

Stories Shared by the First Author

From the academic side, the challenge has often been how to make a usability project both practically useful and academically acceptable. The first author, Rong, entered the UX field from a general user research background. While familiar with the empirical research process, she found, when initially undertaking projects that had a timeline tied to a practical redesign, the demand to meet the stakeholders' needs drove most of the testing and research process. Although she could fulfill the requirements of these projects, much of the work was not considered conceptually valuable from an academic standpoint. The primary challenge she faced was designing projects that could yield scholarly merit while still adhering to the practical timelines and resource constraints typical of product development.

To ensure that practical projects could lead to publishable papers, Rong had to overcome two key issues: ensuring an adequate sample size and the need for findings to be statistically valid and conceptually insightful.

One example of these challenges occurred in a project Rong conducted, which had two phases. In Phase I, 30 participants individually interacted with a Microsoft Surface Table in a science library. While the study identified several concrete areas for improvement and was practically useful, the paper submitted to an academic conference was rejected due to the small sample size and the findings being deemed insufficiently interesting. Despite the project's practical benefits in improving an interactive tabletop app, it did not meet the rigorous standards required for academic research.

To address this issue, a Phase II study was conducted with 60 participants working in pairs. To better understand how interactive tabletops support collaboration, Rong reviewed existing theories and used collaboration models to code the data. The Phase II study was successfully accepted by an academic conference (Tang & Quigley, 2014), but it required significantly more effort than what is typically needed in practical projects. This experience highlighted the difficulty of creating research that is both publishable and able to drive timely, effective improvements in a real-world setting.

On the other hand, it is also clear to the first author that academic theories can support better understanding of user behavior patterns or trends. For instance, in a visit to a software company, Rong learned that in their internationalization effort, they discovered the role cultural norms may play when internationalizing icons. For instance, if a turtle happens to be chosen as an icon, users from North America may consider it lovely, but it would trigger different sentiments in users from China and users from Japan. In Japan, the turtle is viewed as a respected animal with a long life. To a Chinese audience, on the other hand, the turtle might be used to represent lowlife or cuckoldry. Having an academic framework for cultural differences could have helped lead to a better choice in icon design.

Meanwhile, Rong also saw how scholarly theories could support a better understanding of user experience. For example, during a visit to a software company, she learned about their internationalization efforts, which may include choosing icons for different cultures. Cultural norms play a significant role in icon interpretation: while a turtle might be considered cute in North America, it evokes very different feelings in other cultures. In Japan, a turtle symbolizes respect and longevity, but in China, in certain contexts, it might be used to represent lowlife or cuckoldry. An academic framework for understanding cultural differences could have helped the company make better design choices from the start.

Stories Shared by the Second Author

The practical need for UX work drew Graham, the second author of the book, into the field. He learned about that practical need that hard way, by building an interface without doing enough research to determine whether it would be needed. He spent months trying to build an interface to distribute the course evaluations that students completed about their instructors at the end of each semester. Although such an interface seemed like a good idea, it was not practical, as there were privacy concerns related to sharing the reviews online. Also, a similar system was in the process of development by a separate university team. This experience drove him to realize the importance of pursuing research first to uncover users' real needs. The project constraints drove him to understand how crucial UX research is in informing design. To him, the field of UX is a practical way to avoid unnecessary engineering work.

In his professional practice, Graham often finds that the quicker and less scientifically documented a design or research finding is, the more likely it is to succeed in the workplace. When first entering the workforce from school, his design deliverables read like essays. Due to their length, his ideas gained little traction. Engineers and product managers, unlike academics, tend to have neither the time nor the patience to read lengthy explanations of research with all the nuances of the data; succinct summaries of recommended product changes work better in the practical realm, especially when those summaries are presented in a way that allows people to efficiently internalize the data, develop their own practical understandings of the obstacles users experience, and formulate solutions.

On the other hand, Graham also encounters situations where he sees the value of scholarly knowledge in interpreting findings from applied research. Recently, he was sitting around a table designing a webpage when the questions came up: do people understand the three bar "hamburger" menu at the top of sites? What is the best way of presenting a button to open a navigation menu on a cell phone? A button to open a mobile menu is a widespread thing. Surely there must have already been someone somewhere who has studied it. There are probably even papers about the theory of mind behind people's interpretation of navigation, portals, and horizontal lines. However, because Graham didn't know of any recent existing study, he ended up setting up an A/B test to study several variations of a menu to see which attracted the most clicks. It turned out that the "hamburger" was the most clicked variant, but it wasn't clear that people knew what they would get if they clicked it. Having easy access to published research about users' mental models of menus for mobile devices could have facilitated a better understanding of the reasons for certain user behavior patterns to inform the UX design. Graham's experiences show us that an academic foundation has the potential to save both time and resources.

What This Book Is and Is Not

As with any book, this one has specific focal points and does not cover everything in the vast universe of UX. Below, we clarify what this book aims to achieve and what it does not, to set accurate expectations.

What This Book Will Do

This book aims to:

- Explore the concept of UX as defined in academic and applied UX research, and identify the gaps between these definitions in the two settings.
- Examine the methods and techniques used in UX research in both settings, and highlight the differences.
- Analyze the styles of communication and ways of sharing and reporting UX research findings in academic and applied settings, and identifying discrepancies.
- Propose concrete steps to bridge the divide between academic and applied UX. This includes offering a framework for convergence, with the goal of strengthening, advancing, and empowering the UX field and its communities.

What This Book Isn't

It's important to note that this book is not:

- A comprehensive meta-analysis of theories, methods, and empirical findings in UX research since the field's inception in the 1950s (Nielsen, 2017).
- A practical guidebook on UX design or UX research
- A discussion of the specific knowledge, skills, and competencies required for academic UX researchers or professional UX specialists
- A comprehensive textbook that introduces UX topics and outlines stages of UX research processes
- A complete resource for all definitions, methods, and communication strategies of UX research in both academic and applied settings.

References

Lallemand, C., Gronier, G., & Koenig, V. (2015). User experience: A concept without consensus? Exploring practitioners' perspectives through an international survey. *Computers in Human Behavior, 43*, 35–48. https://doi.org/10.1016/j.chb.2014.10.048

Law, E., Roto, V., Hassenzahl, M., Vermeeren, A., & Kort, J. (2009). Understanding, scoping, and defining user experience: A survey approach. In *Proceedings of 2009 ACM CHI*, Boston, MA, 2009, pp. 719–728. https://doi.org/10.1145/1518701.1518813

Nielsen, J. (2017). A 100-year view of user experience. Retrieved from: https://www.nngroup.com/articles/100-years-ux/

Tang, R., & Quigley, E. (2014). Dyadic diversity attributes on interactive tabletop collaboration, performance, and perception. In: Proceedings of iConference 2014. https://www.ideals.illinois.edu/items/47335/bitstreams/138878/data.pdf

Definitions of UX

2

Overview

Since the term User eXperience (UX) was first introduced in the 1970s and gained more widespread use in the 1990s (Norman, 2013), its definition has varied across academic disciplines and professional practices (Alves et al., 2014).

Lallemand et al. (2015) noted that "since the 2000s, the concept of UX is widely used but understood in different ways" (p. 35). Similarly, Law et al. (2009) observed that despite numerous efforts within the UX community to develop a unified definition, "a consensual definition of UX is still lacking" (p. 719). They argued that establishing a universal definition would have three key benefits:

1. It would "facilitate scientific discourse" (p. 719),
2. It would "enable managing practical applications of UX" (p. 720), and
3. It would "help in teaching the notion of UX" (p. 720).

The challenge of deriving a common definition of UX prompted Law and her coauthors to conduct a global survey, aiming to explore the possibility of a unified UX definition. They selected five definitions representing different perspectives, addressing the core elements of UX—who, what, how, and when. A total of 275 participants responded to the survey, conducted before, during, and after the ACM SIGCHI 2008 conference (which typically attracts a more academic audience compared to conferences such as UXPA or IXDA, which appeal more to practitioners). The survey revealed a clear divide: industry respondents preferred a UX definition originating from industry, while academics favored two definitions that were authored by people from their own domain.

Four years later, Lallemand et al. (2013, 2015) replicated Law et al.'s survey, translating it into French and German, and garnered responses from 428 participants.

© The Author(s), under exclusive license to Springer Nature Switzerland AG 2025   7
R. Tang and G. Herrli, *The Domain of UX in Information Studies: Bridging Theories, Research, and Professional Practice*, Synthesis Lectures on Information Concepts, Retrieval, and Services, https://doi.org/10.1007/978-3-031-83529-2_2

Their findings confirmed statistically significant differences in UX definition preferences between academic and industry participants. As in the earlier study, industry professionals leaned toward industry-based definitions, while their "academic counterparts" preferred academic definitions.

Interestingly, when Lallemand et al. compared the results of their survey Law et al.'s (2009), they noticed a shift: while an academic-oriented definition remained the top choice for both studies, the preference for an industry-based definition had grown by more than 8%. The authors attributed this evolution to a larger sample of UX practitioners and the growing popularity of UX, which has broadened its application across diverse contexts. They concluded that the industry-based definition's broad, easy-to-understand nature may appeal to a wider audience, helping to unite various UX stakeholders—"The industry-based definition is an easy-to-understand definition that confers an extremely broad scope to UX and might therefore unite numerous respondents" (p. 45).

Despite numerous efforts to establish a unified definition of UX, a conceptual gap persists between academia and industry, with each side informed and shaped by its own domain-specific perspectives. As Alves et al. (2014) noted, "we sought the literature and failed to find a comprehensive multidisciplinary overview on UX evaluation practice" (p. 94). While "seeking the literature" for a definition is a distinctly academic approach, the elusive nature of a universal UX definition underscores the challenge of bridging these differing viewpoints.

Academics attempting to define UX often face a barrier in that applied practitioners may see little value in such definitions. Kaikkonen (2009), in their dissertation research, highlights this divide: "From the practitioners' point of view, the current definitions have only relative value; they may build the framework in research, but do not offer tools to effectively create or evaluate user experience" (pp. 25–26). In the following sections, we explore academic and industry definitions of UX, pinpoint areas of disconnect, and propose strategies for bridging these seemingly disparate perspectives.

Academic

From the scholarly perspective, the conceptualization of UX is profoundly rooted in psychological theories and empirical research on human experience. It is, therefore, a more acceptable and appealing term than "usability" within academic circles because of its broader scope, encompassing sensation, emotion, and the "non-utilitarian aspects" of human interaction (Law et al., 2009). In Law et al.'s study, academic respondents favored Hassenzahl and Tractinsky's (2006) definition of UX, which emphasizes the psychological factors that shape a user's experience. The following sections provide an overview of the history and evolution of the term UX.

History

History of the Concept of UX

The phrase "user experience" can be traced back to the 1960s to early 1970s in human factors' literature. Early instances, though different from today's understanding, appeared in discussions of user interactions with products. For example, In 1965, Freeberger's abstract for The Evolution of a Pacemaker noted, "Today's military 1/4 ton vehicle has evolved from many years of design, test, production, and user experience." Similarly, Worlton (1971) stated, "From user experience, applications, and the survey results, requirements for two classes of bulk storage devices emerge—dynamic and archival bulk storage devices."

These uses of the term "user experience" appear to refer to the history of using certain devices or tools, but the term UX was not yet distilled into a shorthand abstraction representing an area of study or professional expertise. In fact, when the phrase "user's experience" was used in the human factors or engineering literature, it was obviously referring to the personal history of user in using the product over time (e.g., Gardner, 1981).

In these early examples, "user experience" referred to the accumulated history of using specific devices or tools rather than the distilled abstraction we now associate with the term. When the phrase "user's experience" was used in human factors or engineering literature, it typically referred to a user's personal history with a product over time (e.g., Gardner, 1981). The concept had not yet evolved into a distinct area of study or professional expertise.

One early use of the term "user experience" in a more modern sense can be traced to Edwards and Kasik's (1974) presentation at the VIM-21 conference, titled "User Experience with the CYBER Graphics Terminal." This article has been recognized by several sources as an early example of using UX in the way we understand it today (al-Azzawi, 2014; Knemeyer & Svoboda, n.d.). However, after 1974, the term was not widely used until Norman and Draper's 1986 book, *User Centered System Design,* where they referred to "the user's experience." They stated, "This section of the book contains chapters that get directly at the question of the quality of the user's experience... This is of course the ultimate criterion of User-Centered System Design" (Norman & Draper, 1986, p. 64).

Many scholars (e.g., Alves et al., 2014; Lallemand et al., 2015) credit Donald Norman as one of the first to popularize the term "User Experience." In his 1988 book, *The Psychology of Everyday Things*, Norman (1998) dedicated a chapter to "User-Centered Design," outlining cognitive principles for interface design and discussing the user's mental model alongside the designer's conceptual model. When Norman joined Apple Inc. in 1993, he labeled his division the "User Experience Architect's Office," helping to mainstream the term UX as broader than "usability" and more focused on the overall human experience (https://hci.stanford.edu/publications/bds/12-norman.html).

In a 1995 interview conducted by John Rhinefrank, Norman explained that "I work in a group we call the "'user experience architects office' because we want to emphasize the experience of working with a particular technology, how the experience feels" (Rhinefrank, 1995, p. 51). Twelve years later, in an interview with Peter Merholz (2007) of Adaptive Path, Norman elaborated: "I invented the term because I thought human interface and usability were too narrow. I wanted to cover all aspects of the person's experience with the system including industrial design graphics, the interface, the physical interaction and the manual" (Merholz, 2007).

In late 1994, Apple's Human Interface Group (HIG) was disbanded and replaced by the User Experience Research (UER) group, as noted by Russell (1998). The UER group was created with the specific goal of bridging the gap between design domains—graphical, interaction, and visual—and system implementation. Russell highlights that the UER group emphasized "user studies, speculative design, implementation, and learning from the experiences of deployment" (p. 90). Prior to Apple's UER group, similar groups existed in other companies, such as Compaq, which had a "Human Factors" (HF) group. The HF group at Compaq aimed to "improve the usability of COMPAQ products, allowing end users to accomplish their tasks more effectively, efficiently, and comfortably" (Human Factors Group, 1992, p. 285).

Apple's UER group exemplifies how UX-focused teams were starting to emerge in corporate settings in the early 1990s. In 1997, Netscape established its own user experience team, initially named the "Gooey group," which became one of the fastest-growing design teams in the software industry (Fernandes, 1997). According to Fernandes (1997), the Netscape User Experience Group played a crucial role in product development, focusing on "usability testing, visual design, and interaction design" (p. 88). By the late 1990s and early 2000s, more companies reported the success of their UX teams (e.g., Tripathi, 2007; Levi et al., 2007), further cementing UX as a critical component in corporate product development.

Regardless of who actually "coined" the term "user experience," since the 1990s, UX has gained widespread recognition in both academic and professional circles. However, pinpointing the first complete definition of user experience is challenging. In 1996, the Association of Computing Machinery (ACM) hosted its inaugural ACM/Interaction Design Awards. Rather than establishing judging criteria in advance, the judges derived eight criteria from their evaluation of the submitted products (Rettig, 1996a, b). In describing the criteria related to the quality of experience, Alben (1996) identifies two primary categories: one focuses on "a direct contribution to the user experience," and the other pertains to "the development process used by the product's designers" (p. 13). Alben elaborates: "All the criteria we describe are factors either contributing to or components of the user's experience of the product" (p. 13). Alben further clarifies the concept of "experience:"

By "experience" we mean all the aspects of how people use an interactive product: the way it feels in their hands, how well they understand how it works, how they feel about it while they're using it, how well it serves their purposes, and how well it fits into the entire context in which they are using it. If these experiences are successful and engaging, then they are valuable to users and noteworthy to the interaction design awards jury. We call this "quality of experience." (p. 12)

In reflecting on the awards criteria, Marc Rettig (1996a, b) highlights that "to create a great user experience, a design must be all of these things: Needed... Grounded in Understanding... Learnable and Usable... Appropriate... [the] Product of [an] Effective Process... Aesthetic... Manageable... Mutable" (p. 64). The deliberate effort to establish and publish these eight criteria, along with the candid and passionate exploration of the attributes that define effective interaction design and quality of experience, can be seen as one of the earliest holistic attempts to define "user experience." At the time, the terms "user's experience" and "user experience" appear to have been used interchangeably. Both expressions are found in articles published in the May/June 1996 special issue of Interaction magazine, which was dedicated to the *ACM/Interaction* Design Awards.

While the 1996 *ACM/interaction* Design Award sparked significant discussion around user experience, it did not explicitly define the term "User Experience" (UX). Due to the "heterogeneous interpretations of UX," and its frequent conflation with related terms like Human–Computer Interaction (HCI), usability, and Information Architecture (IA) (Fredheim, 2012), a more targeted conceptualization of UX—its components, theories, and scope—was still needed.

Several research articles have since contributed to advancing the 1996 ACM conceptualization. For example, Kerne (1998), Pine and Gilmore (1998), and Forlizzi and Ford (2000) approached the notion of experience through various lenses, including cultural representations, realms of experience, and the role of interaction designers in understanding user experience. Kerne (1998) sought to expand the formative questions and guidelines by introducing concepts such as interface ecosystems and cultural representation. For instance, in response to the questions about "understanding of users," he suggested incorporating inquiries about the design process and interactions to ensure they "speak in the users' languages" (p. 39). Regarding "aesthetic experience," Kerne (1998) recommended revising the guidelines to include questions such as, "Does the information space offer an open-ended range of possible user experiences?" (p. 40).

Pine and Gilmore (1998) emphasized the pivotal role of experience in the evolution of economic value, arguing that following the shift from an industrial economy to a service economy, we are now entering the "emerging experience economy." They noted that compare to previous economic offerings—commodities, goods, and services—which are external to the buyer, experiences are personal and subjective, existing only "in the mind of an individual who has been engaged on an emotional, physical, intellectual, or even spiritual level" (p. 99). Pine and Gilmore (1998) further proposed categorizing experiences into "four broad categories" based on two dimensions: "participation" and "connection."

The four realms of experience, differentiated by the axes of active versus passive participation and immersion versus absorption. For example, entertainment falls within the realm of "absorption" and "passive participation," while the "escapist" realm combines "active participation" with "immersion."

For Forlizzi and Ford (2000), experience extends beyond the products that are merely "pleasant or easy to use" (p. 421). They propose four components of experience in the context of interaction design: Sub-consciousness, cognition, narrative, and storytelling. Sub-conscious experiences involve routine activities that "do not complete our attention and thinking process," cognitive experiences "require us to think about what we are doing" (p. 421). Meanwhile, narrative experiences are those that have been internalized and formalized in the user's mind through language, whereas storytelling experiences are more subjective, where "a person relays the salient parts of an experience to another, making the experience a personal story." These stories allow users to "bestow meaning on situations, creating life stories and stories of product use" (p. 422).

In the corporate setting, the way UX practice is described often helps define the term itself. Russell (1998) explains that for Apple's User Experience Research (UER) Group, "the idea of user experience is to take care of, account for, design and consider everything the user uses. Creating a useful and enriching user experience is an encompassing goal that necessarily crosses specialty boundaries in the pursuit of a single, unified, coherent experience of the computational" (p. 90). Russell (1998) highlights that the UER group sought to create a new kind of user experience through a multidisciplinary approach. This approach went beyond user interface considerations, incorporating factors such as "the physical design of the system, settings of use, metaphor design, and expectations about the system" (p. 90). He also provided examples of several projects the UER group was involved in, emphasizing the importance of strong organizational commitment to bring these innovations to fruition.

Nevertheless, in reflecting on the challenges, Russell (1998) notes that many of these projects failed to transition from research to production. He attributes this primarily to the significant effort required to integrate these ideas into production, stating, "there is no shortage of great research and concepts—but there is an understandable reluctance to commit many resources to new user interface concepts, especially when so much remains as yet undone in the mainline" (p. 93).

Several years after the ACM special issue on *ACM/interaction* design awards, researchers from the CHI (Human Factors in Computing Systems) and HCI (Human Computer Interactions) communities initiated a series of discussions focusing on the definitions of UX. These discussions took place across various conferences, including NordiCHI, HCI, and Interaction, through workshops, panel discussions, and paper presentations. This collective effort may have contributed to the exponential growth in publications exploring the conceptualization of UX (Kuutti, 2010).

The definition of UX can vary in the context of digital gaming. Started in 2004, a research team from the University of Helsinki began developing a UX measurement framework specifically for games and virtual environments (VEs). Särkelä et al. (2004) proposed that UX should be measured by "perceptual-attentive, cognitive-emotional and motivational constructs."

Sarkela et al. (2004) proposed a UX framework for gaming featuring four key dimensions: physical presence, emotional involvement, situational involvement, and performance competence. This framework has been tested, extended, and reframed over time (e.g., Takatalo et al., 2006; Komulainen et al., 2008; Takatalo & Häkkinen, 2014).

In a 2008 study, Komulainen et al. (2008) examined the psychological constructs of UX in games by analyzing the responses of 267 Finnish gamers. They found that, in addition to the typical UX constructs—cognition, emotion, and motivation—gaming UX also involves perceptual intensiveness, or "focused attention," characterized by immersion and role engagement. The cognitive dimension encompasses the gaming process, including the challenges faced and strategies for overcoming them. The emotional aspect consists of valence and arousal, ranging from relaxation to excitement, and frustration to enjoyment. The motivation factor addresses the purpose of gaming, from passing time and escapism to entertainment and even addiction.

Towards a Common Definition of UX

Since the term UX gained widespread acceptance, researchers from both academia and professional practice have struggled to establish a universally agreed-upon definition. While the attention to "experience" has introduced new dimensions to the concept, the absence of a "commonly agreed definition" has left the "state of affairs within the field of research ... naturally quite unsatisfactory" (Kuutti, 2010, p. 715). According to Kuutti (2010), beginning in 2006, "there is an active subcommunity within HCI working on systematizing and conceptualizing the heterogeneous UX field," and as a result, "some consensus has already emerged" (p. 715).

In 2006, the NordiCHI UX workshop centered on the theme "Theorizing, Qualifying, and Quantifying UX." In the foreword to the workshop proceedings, Law et al. (2006) noted that while "the so-called user experience (UX) movement is gaining ground," there was still a significant gap: "Neither a universal definition of UX nor a cohesive theory of experience yet exists that can inform the HCI community how to practically design for and evaluate UX" (p. iii). Law et al. further emphasized that:

> Theoretically UX is currently incoherent, methodologically UX is not yet mature either. Some critics even argue that non-instrumental needs are too fuzzy, elusive and idiosyncratic to operationalize (i.e. they are simply dismissed as intractable) and that experience and emotion are too ephemeral and complex to measure. Proponents of UX are more optimistic. First, within UX there seems a shared understanding that UX needs to clarify and operationalize constructs to be taken seriously within the context of SE or user-centred design. Second, at least some approaches to UX believe that with a proper definition come valid and reliable measures. The latter requires the integration of the many facets of UX into a more unified view. We reached a point, where the pressing question is no longer whether we need UX or not. We need it and we

must work on a shared understanding of what UX is and how it can be addressed by design, engineering and research. (p. iii)

Echoing Law et al.'s concern over the lack of a coherent view of UX, Jetter and Gerkin (2006) observed, "Although the term "user experience" ... seems to be widely adopted by practitioners and the industry all over the world, there seems to be no scientific consensus on a definition, on the scope, or on a theoretical model of UX" (p. 107). This call for a "unified view" or "shared understanding" reflects a recognition of the conceptual gaps between UX in design, engineering, and research. Close to 20 years later, have these gaps been bridged or is there still a big gulf separating the viewpoints of UX?

In their article "User Experience—Towards a Unified View," Hassenzahl et al. questioned what UX truly is and how it differs from usability. They argued that only when people can successfully differentiate UX from Usability will UX be accepted "as a topic in its own right" (Hassenzahl et al., 2006, p. 1). The authors propose that UX is distinct from usability in three dimensions: it is more holistic, subjective, and positive. They also noted a strong interest in defining and mapping UX, suggesting that a unified model might be possible through a component-based approach.

In 2007, Law and her co-organizers proposed a workshop titled "Towards a UX Manifesto" at the British HCI conference, calling for the creation of a "coherent Manifesto" for the UX field.

The workshop proceedings featured exploratory articles on UX principles and frameworks. In their proposal, Law et al. (2007) invited "researchers, educators, and practitioners" to contribute to developing a UX manifesto that would serve as "a reference model" for future UX work.

They outlined three foundational pillars for the manifesto: principle (the underlying assumptions of UX), policy (the positioning of UX relative to other domains), and plans (strategies for improving UX design and evaluation).

The principle of UX involves clarifying "basic concepts and assumptions," the "structure of UX—components & attributes," and the "process of UX" as it relates to time, tasks, or users. Several articles from the workshop delved into defining UX. For instance, Roto (2007) presented seven definitions of UX and emphasized that few definitions captured the critical distinction that while usability is "a product attribute," "user experience is a personal, subjective feeling about the product" (pp. 31–32). Roto further argued that UX differs from other types of experiences because it (1) involves "a product/service," (2) entails "interacting (or the possibility of interacting) with a system," and (3) occurs "whenever there is interaction with a product, even if the product is not interactive" (p. 32). Additionally, Roto proposed that UX exists along a continuum, from the anticipated UX before interaction, to the immediate experience during interaction, to the overall UX, which extends beyond the interaction itself.

Following this, several articles explored different definitions of UX and examined the preferences of individuals in academia versus those in industry (e.g., Law et al., 2009;

Naumann et al., 2009; Lallemand et al., 2015). Law et al. (2009) initiated this line of research by surveying 275 UX researchers and practitioners, asking them to rate their agreement with 23 UX-related statements and to choose among five UX definitions. Although the study revealed significant differences across countries, the authors found that "no patterns describing how the differences systematically vary with background variables can be derived" (p. 726).

Based on discussions from the CHI'2008 workshop, the authors recommend that UX should "be scoped to products, systems, services, and objects that a person interacts with through a user experience" (p. 727). Regarding the temporal aspect of UX, the survey results revealed that UX can be evaluated during the interaction, but overall UX can be assessed either simultaneously or afterwards. Law et al. (2009) concluded that the evolving ISO 2008 definition of UX aligns with the findings of their study, reflecting a growing consensus within the field.

The various workshops, panels, and publications, as noted by Kuutti (2010), contributed to the eventual consensus on the definition of UX, which was formalized in March 2010 as ISO DIS 9241-210 2.15: "a person's perceptions and responses that result from the use or anticipated use of a product, system or service."

The ISO (2010) definition further clarifies:

Note 1 to entry: User experience includes all the users' emotions, beliefs, preferences, perceptions, physical and psychological responses, behaviours and accomplishments that occur before, during and after use.

Note 2 to entry: User experience is a consequence of brand image, presentation, functionality, system performance, interactive behaviour and assistive capabilities of the interactive system, the user's internal and physical state resulting from prior experiences, attitudes, skills and personality, and the context of use.

Note 3 to entry: Usability, when interpreted from the perspective of the users' personal goals, can include the kind of perceptual and emotional aspects typically associated with user experience. Usability criteria can be used to assess aspects of user experience.

Figure 2.1 presents a timeline highlighting the key workshops, panels, and discussions that contributed to the development of the 2010 ISO definition.

After the publication of the ISO (2010) definitions, discussions about establishing a shared understanding of UX continued. These efforts included the Dagstuhl 2010 Seminar and the creation of the UX White Paper, which noted that "the multidisciplinary nature of UX has led to several definitions of and perspectives on UX, each approaching the concept from a different viewpoint." It emphasized that "existing definitions for user experience range from a psychological to a business perspective and from quality centric to value centric. There is no one definition that suits all perspectives" (User Experience White Paper, 2011, p. 4). The White Paper identified three key perspectives on UX: UX as a phenomenon, UX as a field of study, and UX as practice. While the first two dimensions

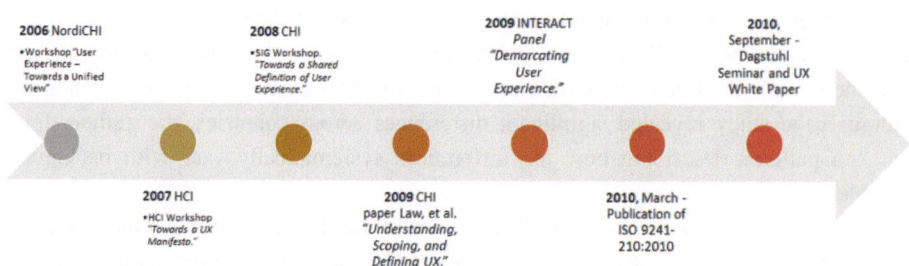

Fig. 2.1 Selected milestone workshops, panels, and discussion on a shared definition of UX leading to ISO (2010) definition

primarily concern academic research, the third is closely tied to practical, applied UX efforts.

Definitions of UX from Researchers in Academia

As documented in earlier sections, the effort to establish a consensual definition of UX has gone through a long journey. As Forlizzi and Battarbee (2004) noted, "The term 'user experience' is associated with a wide range of meanings, and no cohesive theory of experience exists for the design community" (p. 261). Over the years, many definitions of UX have emerged from academia. The following sections outline several influential and widely cited academic definitions of UX.

Definitions by Hassenzahl and His Coauthors

Marc Hassenzahl from Folkwang University has been a prominent voice in shaping the concept of UX. A definition by Hassenzahl and Tractinsky (2006) has been widely regarded as a representative definition within academic circles. They define UX as follows: "UX is a consequence of a user's internal state (predispositions, expectations, needs, motivation, mood, etc.), the characteristics of the designed system (e.g., complexity, purpose, usability, functionality, etc.) and the context (or the environment) within which the interaction occurs (e.g., organisational/social setting, meaningfulness of the activity, voluntariness of use, etc.)" (p. 95).

The definition highlights the importance of users' psychological and personal characteristics in shaping their interaction with a system within a certain context. It has been frequently cited as the preferred definition by researchers (Law et al., 2009; Lallemand et al., 2013, 2015).

In his 2007 UX Manifesto article, Hassenzahl (2007) differentiates UX from Usability by emphasizing the hedonic and subjective aspects of UX:

I believe UX to differ fundamentally from usability, because of its focus on (a) positive aspects of the user-product relationship (e.g., enjoyment rather than frustration), (b) the incorporation of hedonic (non-instrumental) aspects and (c) the focus on the understanding and management of the subjective side of product use (rather than objective performance criteria). (p. 13)

Hassenzahl further explored UX through two levels of goals: "do-goals" (pragmatic quality) and "be-goals" (hedonic quality) (Hassenzahl, 2003, 2008). He argued that UX "encompasses all aspects of interacting with a product," emphasizing that UX is subjective, varies between individuals based on personal standards, and changes across situations and over time (Hassenzahl, 2003, p. 11). He reinforced this in 2008 by defining UX as "a momentary, primarily evaluative feeling (good-bad) while interacting with a product or service" (Hassenzahl, 2008, p. 12). Throughout his work, Hassenzahl consistently highlights the hedonic, subjective, psychological, temporal, personal, and situational dimensions of UX.

Definitions by Desmet and Hekkert

Desmet and Hekkert, from Delft University of Technology in the Netherlands, approached UX through the lens of a product experience framework. According to Desmet and Hekkert (2007), this framework "applies to all affective responses that can be experienced in human-product interaction" (p. 1). Hekkert (2006) provided a more specific definition, stating that UX encompasses "the entire set of effects that is elicited by the interaction between a user and a product, including the degree to which all our senses are gratified (aesthetic experience), the meanings we attach to the product (experience of meaning), and the feelings and emotions that are elicited (emotional experience)" (p. 160).

This definition has been widely recognized and adopted by researchers in academia. The product experience framework, essentially a UX framework, includes three key components: aesthetic experience, experience of meaning, and emotional experience. Figure 2.2 is the framework presented by Desmet and Hekkert (2007).

This definition, like Hassenzahl's, emphasizes the subjective, emotional, and personal aspects of user experience. Nevertheless, it also introduces cognitive experience (experience of meaning) and aesthetic experience (related to design). Desmet and Hekkert (2007)

Fig. 2.2 Framework of product experience. *Source* Desmet and Hekkert (2007)

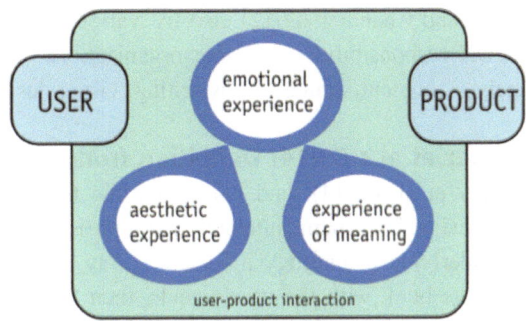

note that the UX discourse within the design research community "draws on an extensive set of affective or experiential concepts" (p. 8). They further explain that "an understanding of affective experience will require an approach that explains how behaviour, cognition, and experience are interrelated in human-product experience" (p. 8).

Definition by Robert and Lesage

In 2011, Robert and Lesage (2011), faculty members from the École Polytechnique de Montréal and Université de Montréal, presented their analysis of the UX concept and proposed following definition: "UX is a multidimensional construct that defines the overall effect over time on the user of interacting with a system and service in a specific context" (p. 311). They further elaborated the characteristics of UX through a comprehensive list of elements. In Fig. 2.3, the first author attempts to include various aspects covered in their explanation of the definition. This definition, along with its detailed explanation of various aspects of the concept of UX, appears all-encompassing, reflecting the academic approach of integrating relevant discussions into a matrix of comprehension of UX as a concept.

Applied

Definitions of UX from Applied Practitioners

Unger and Chandler's definition from A Project Guide to UX Design.

When asked what UX design is, the go-to definition for Graham (one of this book's authors) is Unger and Chandler's (2012). They treat UX as an overarching field that encompasses three main pillars:

1. information architecture (structuring information to be accessible)
2. interaction design (creating an interface that responds in expected ways)
3. user research (learning who the users are, and what they need, and then learning what issues they encounter when using designed solutions).

Unger and Chandler (2012) also list various fields that are related to UX, and that may be UX responsibilities at some organizations. These include: brand strategy, business analysis, content strategy, copywriting, visual design, and front-end development.

Cooper et al.'s (2014) Definition from About Face

Cooper et al. (2014) tend not to use the term UX; instead they write about "interaction design." To them, interaction design is a way of presenting a business model and an underlying technology model in a way that fits within a user's mental model. They open the book with the user's goals, then focus on how to design behaviors to accomplish these goals, and eventually narrow their focus to specific interaction patterns. Their

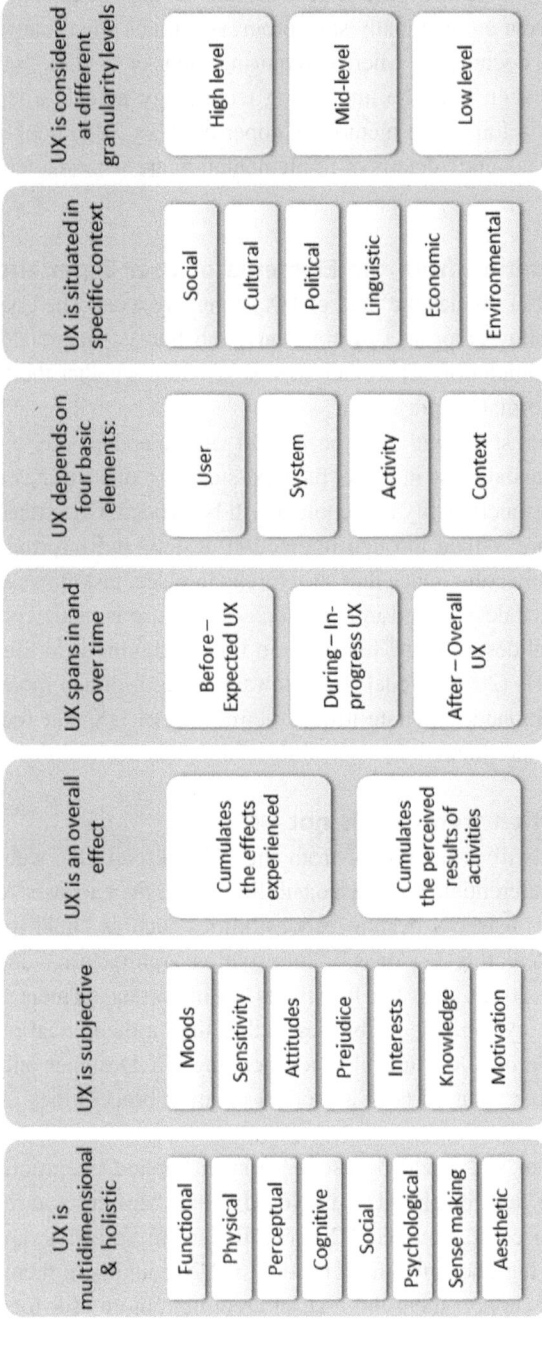

Fig. 2.3 Summary of what makes UX. Created based on Robert and Lesage (2011)

recommended patterns give examples of how to design in ways that fit a user's mental model. For example, they write about the pattern of auto-saving documents with a version history. Having a document automatically save progress mimics what many have come to expect from using paper documents, whereas requiring the user to click "save" forces the user's mental model to align with the underlying technology model, where a computer has to save the changes in long-term memory. Cooper et al.'s (2014) book is full of such concrete details because concrete details of implementation are what matter to the applied practitioner.

Garrett's (2002) Definition from the Elements of User Experience

Jesse James Garrett (2002) divides the field of UX along two axes. The one is abstract to concrete, and roughly follows a project from its early (abstract) conceptual phase through its eventual (concrete) completion. The other axis is based on whether the focus is on the user's task or on the information presented.

In his model, a project starts with user needs and site objectives (i.e. business goals). After that conceptual foundation is in place, the next step is to make things more concrete through specifying what functionality and content will be needed. Once that basic scoping is in place, task flows are refined through interaction design, and information is refined through information architecture. Once that structure is in place, task flows are made more concrete through interface design, and information architecture is made concrete through navigation design. Visual design is the final step to finish bringing it to life.

Every part of Garrett's (2002) model flows toward making things more tangible and concrete. Applied practitioners gravitate toward definitions of UX that focus on how to turn things into a usable product.

Erik Flowers's Definition from "UX Is not UI"

Some UX practitioners will print posters from the "UX is Not UI" website (Flowers, 2012) in an attempt to differentiate their responsibilities from the narrower responsibilities of UI design. The poster lists UX design responsibilities such as "field research," "face to face interviewing," "creating personas," "information architecture," and "brainstorm coordination" (Flowers, 2012). The list highlights an important element of how active practitioners define UX: by job responsibilities, rather than a theoretical purpose for the field. A unified definition of UX is useful to a practicing UX Designer only insofar as it enables them to determine what parts of a product-creation process they are responsible for.

A unified definition of UX could be useful to such applied practitioners insofar as it helps support the last item on the "UX is Not UI" list: "design culture evangelism." This item points to a need for applied UX practitioners to spread awareness of their job responsibilities. Having one standard definition of UX could help them to own those responsibilities and thus support their end goal of creating a more user-focused product.

Job Responsibilities that May Fall Under the UX Umbrella

Perhaps the best place to gain an understanding of how UX responsibilities are divided in their practical application are UX job listings. A perfunctory review of open positions posted on sites such as indeed.com and simplyhired.com. Reveals a wide variety of responsibilities associated with UX, and shows a remarkable degree of inconsistency in what responsibilities fall under what titles.

User Research

In the 2018 UXPA Salary Survey (UXPA, 2018, p. 12), 1326 respondents were able to select multiple job titles. The most commonly selected title was User Researcher, which was chosen by 56% of respondents. The very common title of UX Designer was not an option in the survey, which casts some doubt as to the reliability of the titles listed. The 2016 O'Reilly Design Salary Survey (King & Magoulas, 2016, pp. 1, 9, 10) instead collected job title as an open text field. Their respondents entered 183 unique titles in 324 total responses. The responses were then tagged. The most common tags were UI/UX (22%) and Designer (22%). The number of respondents who described their jobs as "user research" or "usability testing" was less than half as many as those who said they did "user interface design."

UX work is commonly applied at two very different stages of a project: at the beginning, in order to identify user needs and define a product purpose, and then later, once a designed solution exists, in order to evaluate how well that design solves problems. Both stages together are commonly the responsibility of one person in the role of UX Researcher or UX Designer, although they may be broken out into separate roles in larger companies. One person engaging in two very different stages of the product development lifecycle can be challenging to communicate when beginning to integrate a UX role at an organization and somewhat counterintuitive to grasp when transitioning to practicing UX from an academic grounding of UX.

Needs Gathering (formative/attitudinal research)
The initial needs gathering elements of UX research are commonly rooted in ethnographic methods from social sciences such as psychology and anthropology. In some large companies, this qualitative needs-gathering process may be broken out as a separate role called a Qualitative UX Researcher or Ethnographic User Researcher. Other job titles apply to roles centered upon more quantitative data. A Market Researcher uses analytics data to validate large-scale trends, a title that blurs into Business Analyst and Data Scientist, both of which also construct models on large data sets to try to explain market trends and predict the viability of products.

Problem Finding (summative/behavioral research)
One of the best-known types of user research is the usability study, which involves asking participants to complete tasks using a product and seeing where they run into problems. In large companies, the responsibility of conducting usability studies may be broken out as its own role, called Usability Analyst or Human Factors Engineer. These studies also may be covered by the same person doing the interaction design, in which case the role is often called UI Designer, UX Designer, or Interaction Designer.

Interaction Design

The most common responsibilities of respondents to the 2016 O'Reilly Design Salary Survey were "user interface design," "wireframing," "sketching," and "prototyping" (King & Magoulas, 2016). (This might be taken to imply that these are the most common responsibilities for people in the UX Design field, but there could also be some selection bias in who responded to that particular survey.)

Interaction design is also a job responsibility that may be divided multiple ways. In some workplaces, a UX Designer will handle big-picture workflow (wireframes, prototypes, user flows, and site maps) while an Interaction Designer will handle details of interaction within a webpage, such as visual styles, animations, and UI patterns. In other words, UX Designers focus on big-picture work, whereas Interaction Designers focus on vital details.

In other organizations, the Interaction Designer performs big-picture work, while the details are seen to by a Graphic Designer, Visual Designer, or Interface Designer (which are three different names for nearly identical roles). This lack of consistency in what responsibilities fall under the role of "Interaction Designer" illuminates the need for greater academic unification of the field. The job responsibility of "Interaction Designer" is loosely connected to the academic field of "Human Computer Interaction," although that term is rarely, if ever, used by applied practitioners. Greater consistency in defining what responsibilities fall under what title could help enable job seekers to find roles that fit their interests. It could also facilitate employers' recruitment, making it easier to post a job with a title that best aligns with the skillset of the new hire needed.

Information/UX Architecture

Information Architect is a fairly uncommon job title, perhaps in part because few websites or apps have a sufficiently complex information structure to merit hiring a full-time employee dedicated to the role. Frequently, information architecture is one of several responsibilities of a more generalized UX Designer.

Information architecture ranks as the seventh most common responsibility in the O'Reilly 2016 survey (with 70–80% of the number of responses of the top three responsibilities) meaning that it is commonly the responsibility of UX design professionals, even though only 5% of respondents to that survey had the word "Architect" in their job title, and less than 2% had "Information" (King & Magoulas, 2016).

Related responsibilities (e.g., branding, content management, web development).

Various other occupational responsibilities are related to UX. These responsibilities are sometimes directly the responsibility of someone with a UX job title. Other times, they're handled by individuals without UX experience, potentially in collaboration with a formally designated UX role.

One such responsibility related to UX is Project Management, which was listed as a responsibility by 37 respondents of the O'Reilly 2016 survey (compared with 71 responses for the most common responsibility) (King & Magoulas, 2016). In other words, in building a user-centered product, sometimes project management responsibilities (such as scoping a product and coordinating engineering effort) are owned by the same person who uncovers the user need, and sometimes they're separate roles. The same is true for roles such as UX Developer, which combine front-end coding with interaction design. (To ideally serve the user, of course, the user's mental model and the underlying implementation model should be separated. But in the practical reality, a single person often ends up handling both, which may lead to elements of the implementation model bleeding into the interface presented to the user.) Viewed holistically, the roles might fall on a diagram as follows (Fig. 2.4).

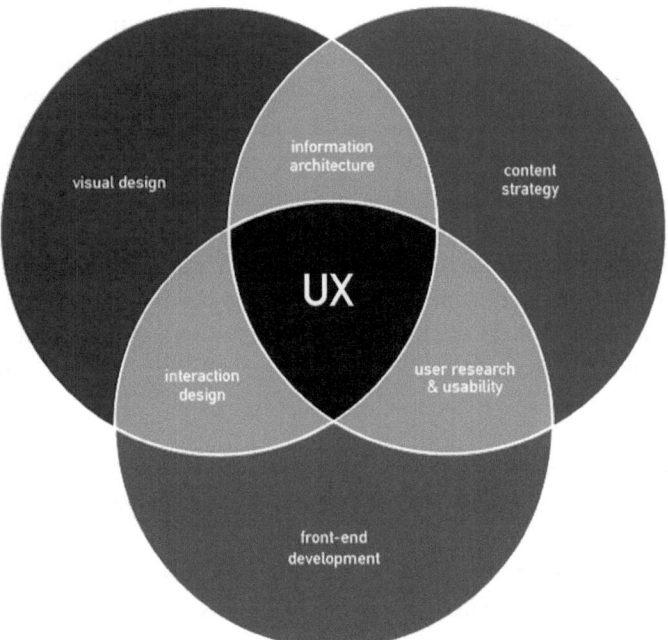

Fig. 2.4 Various UX roles. *Source* http://www.lmbwebdesign.com/wp-content/uploads/2013/09/sta tic.squarespace.com_.png

Gaps Between Academic and Applied Definitions and Conceptualizations of UX

Overview

As discussed earlier, distinct definitions of UX emerge from academic and industry contexts. Dray (2009), then Director of Publications of UPA (Usability Professional Association), highlighted the differing incentives and pressures between these two realms. In academia, *publish or perish* mantra prevails, where "the most basic measures of success are typically scholarly publishing and obtaining grants" (p. 2). In contrast, practitioners in industry operate under the principle of "produce or perish," focusing on learning about "users, users' context and the usability of products" to support business goals, "rather than to satisfy an intellectual interest" (p. 4). Norman (2010) also emphasized that "the gap between research and practice is fundamental." as the required knowledge and skills for each group are inherently different (p. 9).

Both Dray (2009) and Norman (2010) stressed the importance of bridging the gap. Norman (2010) suggested through "translational development," while Dray (2009) advocated for "an infusion into industry the type of critical thinking associated with academia, and of practice that is backed up by sound research" (p. 6). Dray (2009) further asserted that building "alliances" between academia and industry would benefit both sides immensely.

Also highlighted earlier, numerous research studies have explored the differences between academic and industry definitions of UX (Law et al., 2009; Naumann et al., 2009; Lallemand et al., 2015). Both Law et al. and Lallemand et al. found that individuals tended to UX definitions aligned with their own professional backgrounds—academics favoring academic definitions and professionals preferring those from practice. Nevertheless, Naumann et al. (2009) revealed a key distinction in the primary interests of these two groups. For industry participants, the main goal of UX was "to design better products," whereas respondents from academic settings prioritized making "people happier" through their UX efforts. The following sections delve deeper into the conceptual differences between academic and applied UX.

The Academic Need to Establish UX as Its Own Field Versus the Applied Need to Make UX a Part of What Everyone Does

Academically, UX lacks a formal, unified, and certified status, as a departmentally housed distinct discipline. For computer science students, UX is an area of interest within HCI (Human Computer Interaction). For psychology students, it is an applied form of behavioral analysis. For visual design students, it connects with aesthetic principles.

However, a significant disconnect exists among various academic disciplines, making it rare for academic programs to integrate UX into a cohesive, interdisciplinary framework. This lack of a clear academic pathway poses challenges for applied UX practice. Professionals entering the UX field typically come from diverse educational backgrounds, such as psychology, anthropology, computer science, information science, and HCI (Avila, 2017). While there are programs like Bentley University's Human Factors and Information Design Masters that take a holistic approach, a common path for UX professionals to enter the field remains elusive.

In late 2010s, several iSchool programs, including those at UC Berkeley and the University of Washington, have begun offering UX courses and listing UX design as a potential career path for graduates. Special funding has also emerged to support UX education. For example, the University of Tennessee School of Information Science received IMLS funding in April 2016 to develop the Experience Assessment (UX-A) program. Led by Drs. Carol Tenopir, Dania Bilal, and Rachel Fleming-May, the UX-A program "leverages an interdisciplinary team and robust facilities to provide education and experience to future leaders in assessment and user experience (UX) testing" (https://scholar.cci.utk.edu/ux-a/). The program's first cohort graduated in 2018.

As discussed earlier, the lack of a unified definition of UX may have contributed to the limited number of academic programs offering certified degrees specially in UX. Although many LIS programs and iSchools (Thrift & Tang, 2018), very few provide an independent specialization or degree in the field. This absence presents challenges for applied professionals when evaluating the competencies of entry-level candidates. Without a recognized accreditation or clear educational pathway, it can be difficult to assess whether applicants possess the necessary competencies or skills for UX roles.

In response to this gap, various certification programs have emerged, including General Assembly's User Experience Design course, the Nielsen Norman Group's User Experience Certification Program, Udemy's User Experience Certification, the American Graphics Institute's UX Certificate, and Coursera's User Experience Research and Design specialization. For example, Coursera's UX specialization, offered by the University of Michigan's iSchool, consists of six courses, culminating in a capstone project where students design a product from scratch (https://www.coursera.org/learn/user-experience-capstone). However, as these certifications are not issued by accredited bodies within formal educational institutions, their value in the job market remains uncertain. Establishing a standardized curriculum for UX studies, along with core competencies taught across programs, would benefit both current UX practitioners and new professionals entering the field.

UX course offerings from formal higher educational institutions are scattered across various academic departments, with significant differences in how UX is taught in Library and Information Science (LIS) programs compared to Computer Science (CS), business, design, and engineering programs. Moreover, UX instructors, coming from diverse disciplines, do not necessarily have a unified community with consistent communication channels. Currently, no associations or conferences specifically focus on the academic study of UX, limiting opportunities for in-depth exploration, discussion, and advancement of the field.

While UXPA conferences are industry-oriented and focused on best practices, this focus tends to distance them from academic research. UX scholarly work is instead presented at research-oriented conferences including ACM CHI, ASIS&T Annual Meetings, iConferences, and various business, design, and engineering (such as IEEE). However, since UX research is often not a central theme of these conferences, the dissemination of findings and the establishment of UX as a distinct academic discipline remain fragmented. Overall, the community of practice for UX teaching, learning, and research is unformulated and dispersed, hindering the development of a cohesive academic field.

Despite the need to unify the definition of UX, there is simultaneously a very practical imperative to make UX something that everyone does. As Buley (2013) states, when explaining how to make proto-personas with a cross-departmental team: "once you have everyone assembled...help them understand that UX is a frame of mind, and this activity will help get them in that frame of mind. Explain that what you are about to do is an *unscientific* technique that will nevertheless get everyone thinking differently about the people who use your products" (p. 134).

Such an explanation highlights the importance of fostering a collective UX mindset, even if formal definitions remain fragmented.

In an applied setting, treating UX as its own, separate field is counterproductive because knowledge of the user's needs and interests should inform the work of people throughout the organization. Editors (who produce content for users), engineers and designers (who build site features for users), and marketing professionals (who seek to engage a membership of users) all benefit from a greater knowledge of who the users are and what they want. Greater involvement in UX-related activities can ensure that they are able to do their own jobs better by better fulfilling user needs. In this sense, the "UX person" serves as a hub for user-related knowledge. They evangelize for user-centered design practices, seeking to make UX a part of everything rather than its own, distinct field.

Challenges in Transferring Study Findings Outside of Their Original Context

Transferring study results beyond their original context poses several challenges. In particular, applying scholarly concepts to practical settings can be difficult, as can synthesizing applied findings into meaningful academic insights.

Differences in Audience
Differences in culture

Various demographic and psychographic attributes correlate with differences in how people perceive things and assign value to them. Such attributes include ordinary census-type attributes such as race, gender, and household income, as well as more experience-based attributes, such as previous usage of similar products, past addiction, health, hobbies, and interests. Findings from a study of people living an affluent lifestyle may not apply to those of people with more limited finances. Findings from a study of people familiar with one product may not apply to people familiar with another.

Changing perceptions over time

Even if two studies share similar audiences, they may have been run far enough apart in time that those audiences' ideas have changed. For example, a study of how many people interpret a particular icon as representing a way to add a hyperlink may change over time.

In the "Introduction" Chapter, we used the example of familiarity with the three-bar "hamburger" menu, commonly used as a way to open the navigation on mobile devices. One of the authors, Graham, knew of the results of an A/B test that had been published online 6 years earlier, which found that people were more likely to click the word "Menu" than the three-bar "hamburger" symbol. But because usage of that symbol to mean "menu" has become widespread, it was unclear whether people would still overlook it. In replicating the A/B test, Graham obtained results directly as opposed to those of the earlier study. This change might have been a result of the elapsed time (causing more people to be familiar with the symbol) or of the site on which he ran the study having a different audience from the original study (potentially having participants who had been using mobile devices for a longer time).

Another example of how user preferences evolve over time, as well as how the audience itself changes, comes from Rong's observation while leading her UX class in a usability assessment of the Astrophysics Data System (ADS). Initially, many experienced ADS users preferred the "ADS Classic" interface over the newer, more modern design with enhanced visualizations. Despite the Classic interface being form-based, similar to a command language interface, these users valued its precision and efficiency, especially when using pre-scripted search statements. The newer interface left them feeling disoriented.

However, after a few years of acclimating to the modern interface (ADS 2.0, ADS Bumblebee), users grew more comfortable with its updated look and feel. After several rounds of usability evaluations and user research, the designers ultimately decided to phase out ADS Classic. As Accomazzi et al. (2018) explained, "This new search system, which has been in beta for the past three years, has now matured to the point that it provides feature and content parity with ADS Classic, and has become the recommended way to access ADS content and services. Following a successful transition to Bumblebee, the use of ADS Classic will be discouraged starting in 2018 and phased out in 2019." In 2011, during Rong's initial class evaluation of ADS, participants favored the Classic interface. However, by 2018, when Rong's class reassessed the Bumblebee interface, almost none of the participants expressed a preference for the Classic design. The takeaway from this story is clear: as technology evolves, so do users. Insights gathered years ago may no longer reflect current user preferences or behaviors.

Differences in Context

User experiences are highly context-dependent. The context in which a product is used significantly shapes how users interact with it and how they perceive its value. A product that works well for one audience in a particular setting might not be as effective for another audience in a different context. The variability in context makes it critical to tailor UX evaluations to the specific scenarios in which a product will be used. Consequently, findings from one study may not fully be translatable to a different context, even if the product and target audience appear similar.

Differences in interface or product

Products with distinct interfaces or purposes naturally would result in differing user experiences. Findings from one product, especially if it serves a different function or appeals to a different user population, may not easily transfer to another. For instance, a product designed for medical professionals may prioritize functionality and precision, while a product aimed at consumers might focus on its interface's visual appeal and user-friendliness.

Differences in information and how it's presented

Two products that perform similar functions might differ significantly in how they organize and present information. The information hierarchy, or the way content is structured, can greatly influence a user's experience. Even minor adjustments in how information is categorized, displayed, or accessed may lead to very different outcomes. Differences in visual hierarchy, including which elements are given prominence, the color scheme, or typography may also impact how users interpret and engage with information. Overall, even when two products offer similar features, variations in how information is presented can affect usability and user satisfaction.

Differences in triggers and motivation

Users' reasons for engaging with a product may differ in different situations. A person buying shoes for their kid will be different from someone buying shoes for themselves. A study of someone given a gift card to buy shoes on a live site will be different from a study of a person asked to pretend they're shoe-shopping on a wireframe. An individual's motivations, decision-making processes, and user experiences vary. Depending on their motivations and driving factors, an individual's experience with a singular product may change from one interaction to another.

Reliability of Methodology

In academic and applied UX research, there are frequent discrepancies in methodologies used to gather insights, which can hinder the consistency and reliability of findings. As being outlined in the next chapter, differences in research settings, sample sizes, data collection methods, and analytical frameworks may lead to varied results. The methodologies used in academic research, often rigorous and methodical, may not always translate well into real-world applications where faster, more iterative methods are often preferred.

The differences in approaches or methodology underscore the difficulty in aggregating academic insights into meaningful, actionable findings for practitioners, as well as applying practical findings from industry back into scholarly discourse. A more substantial discussion on the methodological or other gaps are presented in later chapters.

The Need to Translate Academic Research into Usable Improvement Plans in the UX Field

One of the major challenges for academic researchers conducting empirical investigations with practical implications is translating their statistically validated results and qualitative observation into actionable design recommendations. As Norman (2010) pointed out, this process requires "translational development" to effectively reinterpret research findings into concrete design proposals that can drive improvements in real-world applications.

In academia, the goal of research—whether pure or evaluative—is often to obtain a large enough sample to compare the performance of various groups and control conditions. However, in the practical realm of UX, specific solutions are needed to inform design decisions and guide revisions. For example, in a study led by Rong, the empirical finding reveals that users familiar with a system outperform those who have never used it, or that individuals with a background in Library and Information Science (LIS) had a more positive interaction with a site like PBCore.org than those without LIS training (Tang et al., 2018). While such scholarly insights are valuable, what matters most to the PBCore.org's developers were the wireframes presented by the researchers on homepage redesign and the new configuration visualization of the PBCore metadata schema.

Academic scholars are often required to yield results that are applicable to a wide variety of systems and products. Focusing too narrowly on a specific system or product can limit the potential for publication, as research must yield insights that transcend individual cases. As a result, academic research methodologies are designed to derive transferable observations that can be applied across multiple contexts.

For instance, the results of an academic UX study on the Astrophysics Data System (ADS) and PBCore.org had broad implications that addressed users' cognitive styles and their eye movement patterns (Tang & Song, 2018). These studies, while presumably specific in their methodologies, suggested design strategies that are applicable to many technical websites (Tang et al., 2018).

For example, a study on the Astrophysics Data System (ADS) (Tang & Song, 2018) and a usability study of PBCore.org (Tang et al., 2018) yielded insights into users' cognitive styles and eye movement patterns. While these studies were specific in their methodologies, the design strategies they suggested had wider applicability, offering valuable guidance for improving user experience across a range of technical websites. Similarly, a usability evaluation on RDMLA (Research Data Management Librarian Academy) learning platform (Tang et al., 2021) not only directly improved the RDMLA interface led the improvement of RDMLA, but also contributed to the pedagogical usability framework by incorporating the "diverse learner population" factor and its attributes. This enrichment to the pedagogical usability concept is the scholarly merit of the study. The capacity to generate widely applicable insights is a critical aspect of academic research, even when studies are rooted in specific systems.

Meanwhile, academic theorists tend to develop grounded theories of user behavior, offering a macro-level holistic view of systems model, while applied practitioners focus on an individual product. In other words, academics aim to explain phenomena, processes, and entire ecosystem, applied practitioners concentrate mostly on product design. One perspective sees the forest, while the other focuses on a single tree.

With this broader "forest-level" vision, academic publications provide valuable insights into large-scale best practices, enriching the industry's understanding of user behavior. Colbert et al. (2005) found that HR managers who engaged with academic research were more likely support best practices derived from it. This finding can be extended to the UX field: the more industry professionals are aware of academic research, the more they align their work with evidence-based best practices. However, as Dray (2009) points out, many academics face challenges in producing research that is relevant and meaningful to practitioners. At the same time, "practitioners tend to see academic research as not relevant to them in reinforcing their professional identity, skills, and legitimacy as the organizational experts in user research" (p. 1). Dray's argument highlights ongoing impediments to the widespread consumption and application of academic UX research within industry settings.

The first major impediment is that most professionals in the field rarely, if ever, read academic research (Jarrett, 2007; Dray, 2009). As Jarrett (2007) explains, the preferences for reading any "detective story" over "nearly any research paper" during their leisure time is obvious, as academic research "generally has to have a business purpose—'help me now!'" (p. 1). The challenge is that most academic studies do not provide the timely, succinct, actionable insights that industry professionals need to address immediate problems.

UX professionals, especially those working in fast-paced, industry settings, often face significant time pressures. Their personal career success depends on their perceived contribution to the product pipeline and their ability to enhance commercial success, making them "indispensable" to the process (Dray, 2009, p. 4). This intense pressure leaves little room for practitioners to delve into academic literature. Jarrett (2007) identifies two primary reasons why industry professionals struggle with academic research: (1) they often cannot understand it, and (2) they cannot easily apply it to their work. The barriers to comprehension, according to Jarrett, begin with the title and extend through the abstract and conclusion, creating a "domino effect of misunderstandings" (p. 3). Jarrett humorously notes, "I am afraid that research is lost to me until some kindly academic colleague decides that it is worth translating into plain language" (p. 3).

More importantly, Jarrett observes that many academic studies rely on university student participants, a group not representative of their stakeholders, which include "the whole adult population, seniors included" (p. 3). Jarrett suggests that academic studies should clearly state their participant demographics early on, rather than making broad generalizations about the applicability of their findings.

Moreover, student participants, frequently the subjects of academic UX studies, do not represent greater populations of users. Jarrett (2007) then described three possible scenarios where valuable academic research may be inapplicable to her, as a UX industry practitioner. She named three ways that academic research is generally inapplicable to industry practitioners: (1) inapplicable because she couldn't understand it; (2) inapplicable because of "an apparent mismatch between their assumptions and the claims that they were making" (p. 5); and (3) inapplicable because "the users, tasks, and context of use are too specialised" (p. 5). Furthermore, Jarrett (2007) identifies three ways academic research is often inapplicable to industry practitioners: (1) it's incomprehensible, (2) there's a mismatch between the researchers' assumptions and their claims, and (3) the users, tasks, and contexts studied are too specialized to be relevant in everyday practice (p. 5). She concludes with practical advice: "If you want your research to be useful to practitioners, make it easy for us to understand, and try to research combinations of users, tasks, and contexts of use that are likely to exist in everyday life" (p. 5).

Other challenges in translating academic research to industry lie in the gaps between readership and study contexts, as outlined in earlier sections. These barriers highlight the broader disconnect between academic findings and the everyday needs of UX practitioners.

The Challenge and Need to Aggregate and Synthesize Applied Best Practices into Conceptually Meaningful Scholarly Understanding

It is challenging for UX practitioners to understand scholarly literature and apply the findings to their practice. It is equally challenging, if not more challenging, for applied UX researchers to aggregate their research discoveries, and synthesize years of UX work, to then make their findings available in the scholarly dialogue. Consequently, whichever discoveries and insights UX practitioners gain from their day-to-day work remain in the practice realm, confined to a particular moment of product design. Any higher-level conceptualization or aggregated thinking tends to be lost, as the applied researcher moves on with their next design project.

The Challenge Among Applied Practitioners

At the time of this writing, a search on LinkedIn.com for current job openings matching "UX" returns more than 68,000 results, and close to 14,500 positions for the term "usability.". A search for people whose profile matches "UX" brings up over 2,000,000, including 548,000 in the United States. That means that there are over a million people distilling insights about their users. Yet those insights are not transferable to other settings. A designer for an ecommerce site selling shoes to middle aged white women may uncover something that another designer for a similar site with a similar audience could use…but there's no system or mechanism in place to enable the transfer of results from one context to the other. Furthermore, as articulated earlier, since the research discoveries that take place in one setting tend not to be readily comprehensible or applicable to another setting with a different set of incentives and pressures, these findings are mostly retained within their own realms.

The same factors outlined above that prevent the translation of academic research into solutions applicable to an applied setting also impede the transfer of applied results to scholarly discourse. In large part, this is due to the sheer number of ways in which studies can differ.

If one UX person's study differed from another's along 20 binary attributes, then that would mean that even if each of the million UX people worldwide contributed a study to a shared and searchable database, many of them would find no study that sufficiently overlapped with their own ($2^{20} > 1,000,000$). Yet, the above lists show that there are easily more than 20 ways in which contexts can differ (there are more than 20 demographic or psychographic attributes of audiences alone!), and the ways in which they differ are rarely limited to just two ways.

The Need to Share Nevertheless and the Great Effort to Cross Over

In spite of all of the above, there still may be some opportunity for applied findings to bubble up to broader academic discourse as well as for empirical findings to be developed into practically acceptable recommendations. Applied practitioners continually satisfice to save time. While a study from one situation may not be fully applicable to another, sometimes it could just be close enough that an applied practitioner may be comfortable using the results of the earlier study (from a slightly different audience with a slightly different context) and not having to invest the effort into re-running the study as a whole. As such, there is some opportunity to aggregate and synthesize applied best practices into conceptually meaningful scholarly understanding.

Meanwhile, responsible academic researchers consistently ask themselves whether their scholarly work can pass the "so-what" test. Driven by the desire to impact real-life situations and make meaningful changes, authors/researchers from academia are often called upon to make their research more accessible. *Atlas of Science* (https://atlasofsc ience.org/about/) or Kudos are just some examples of the effort of making science more understandable by asking authors to use plain or layman's language to summarize their discoveries. Regardless of the challenges in bridging two different settings, it is important to bear in mind that such a need and effort are essential and necessary for the advancement of the UX field.

References

Accomazzi, A., Kurtz, M. J., Henneken, E., Grant, C. S., Thompson, D. M., Chyla, R., ... & Lockhart, K. E. (2018, January). ADS Bumblebee comes of age. In American Astronomical Society Meeting Abstracts# 231 (Vol. 231, pp. 362–17).https://ui.adsabs.harvard.edu/abs/2018AAS...231 36217A/abstract

al-Azzawi, A. (2014). Theories of experience. In: Experience with Technology. SpringerBriefs in Computer Science. Springer, London. https://doi.org/10.1007/978-1-4471-5397-9_2

Alben, L. (1996). Quality of experience: Defining the criteria for effective interaction design. *Interactions, 3*(3), 11–15.

Alves, R., Valente, P., & Nunes, N. J. (2014). The state of user experience evaluation practice. In *Proceedings of the 8th Nordic Conference on Human-Computer Interaction* (pp. 93–102). https://doi.org/10.1145/2639189.2641208

Avila, M. (2017, January 23). *College majors for UX design.* UX Collective. https://uxdesign.cc/col lege-majors-for-ux-design-4a8015fae716

Buley, L. (2013). *The user experience team of one: A research and design survival guide.* Rosenfeld Media.

Colbert, A., Rynes, S., & Brown, K. (2005). Who believes us? Understanding managers' agreement with human resource research findings. *The Journal of Applied Behavioral Science, 41*(3), 304–325. https://doi.org/10.1177/0021886305275799

Cooper, A., Reimann, R., Cronin, D., & Noessel, C. (2014). *About face: The essentials of interaction design* (4th ed.). Wiley.

Desmet, P., & Hekkert, P. (2007). Framework of product experience. *International Journal of Design, 1*(1), 57–66. https://www.researchgate.net/publication/235700959_Framework_of_Pro duct_Experience

Dray, S. M. (2009). Engaged scholars, thoughtful practitioners: The interdependence of academics and practitioners in user-centered design and usability. *Journal of Usability Studies, 5*(1), 1–7. https://uxpajournal.org/engaged-scholars-thoughtful-practitioners-the-interdependence-of-aca demics-and-practitioners-in-user-centered-design-and-usability/

Edwards, E. C., & Kasik, D. J. (1974). User experience with the CYBER graphics terminal. In *Proceedings of VIM-21* (pp. 284–286).

Fernandes, T. (1997). The founding of the Netscape User Experience Group. In *SIG CHI Conference Proceedings* (pp. 87–88).

Flowers, E. (2012, December 15). *UX is not UI.* Hello Erik. http://www.helloerik.com/ux-is-not-ui

Forlizzi, J., & Battarbee, K. (2004). Understanding experience in interactive systems. In *Proceedings of DIS04: Designing Interactive Systems: Processes, Practices, Methods, & Techniques* (pp. 261–268). https://doi.org/10.1145/1013115.1013152

Forlizzi, J., & Ford, S. (2000, August). The building blocks of experience: An early framework for interaction designers. In *Proceedings of the 3rd conference on Designing interactive systems: processes, practices, methods, and techniques* (pp. 419–423).

Fredheim, H. (2012). Why user experience cannot be designed, in Smashing Magazine, *User experience design*. Smashing Media GmbH.

Gardner, P. C. (1981). A system for the automated office environment. *IBM Systems Journal, 20*(3), 321–345.

Garrett, J. J. (2002). *The elements of user experience: User-centered design for the web*. Peachpit.

Hassenzahl, M. (2003). The thing and I: understanding the relationship between user and product. In M. Blythe, C. Overbeeke, A. F. Monk, & P. C. Wright (Eds.), *Funology: From usability to enjoyment*(pp. 31–42). Dordrecht: Kluwer.

Hassenzahl, M., Law, E. L. C., & Hvannberg, E. T. (2006). User experience—Towards a unified view. In *Proceedings of the 2nd COST294-MAUSE International Open Workshop* (pp. 1–3). https://www.irit.fr/recherches/ICS/projects/cost294/upload/408.pdf

Hassenzahl, M., & Tractinsky, N. (2006). User experience—A research agenda. *Behaviour & Information Technology, 25*(2), 91–97. https://doi.org/10.1080/01449290500330331

Hassenzahl, M. (2007). The hedonic/pragmatic model of user experience. In *Proceedings of the 21st BCS HCI Group Conference: Towards a UX Manifesto Workshop* (pp. 10–14).

Hassenzahl, M. (2008). User experience (UX): Towards an experiential perspective on product quality. In *Proceedings of the 20th Conference on l'Interaction Homme-Machine* (pp. 11–15). https://doi.org/10.1145/1512714.1512717

Hekkert, P. (2006). Design aesthetics: Principles of pleasure in design. *Psychology Science, 48*(2), 157–172.

Human Factors Group. (1992). The human factors group at Compaq Computer Corporation. In *Proceedings of the SIGCHI Conference on Human Factors in Computing Systems (CHI '92)* (pp. 285–286). Association for Computing Machinery, New York, NY, USA. https://doi.org/10.1145/142750.142815

ISO DIS 9241-210:2010. *Ergonomics of human system interaction—Part 210: Human-centered design for interactive systems (formerly known as 13407)*. International Standardization Organization (ISO). https://www.sis.se/api/document/preview/912053/

Jarrett, C. (2007). Problems and joys of reading research papers for practitioner purposes. *Journal of Usability Studies, 3*(1), 1–6. https://doi.org/10.5555/2835557.2835558

Jetter, H.-C., & Gerken, J. (2006). A simplified model of user experience for practical application. In *Proceedings of the 2nd COST294-MAUSE International Open Workshop "User eXperience - Towards a unified view" (NordiCHI 2006)* (pp. 106–111). Oslo: The COST294-MAUSE Network.

Kaikkonen, A. (2009) *Internet on mobiles: Evolution of usability and user experience* (Publication No. 200) [Doctoral dissertation, Helsinki University of Technology, Espoo]. TKK Dissertations.

Kerne, A. (1998). Cultural representation in interface ecosystems: Amendments to the *ACM/ interactions* Design Awards criteria. *Interactions, 5*(1), 37–43. https://doi.org/10.1145/268986.268991

King, J., & Magoulas, R. (2016). *Design salary survey: Tools, trends, titles, what pays (and what doesn't) for design professionals*. O'Reilly Media. https://www.oreilly.com/design/free/files/2016-design-salary-survey-report.pdf

Komulainen, J., Takatalo, J., Lehtonen, M., & Nyman, G. (2008, October). Psychologically structured approach to user experience in games. In *Proceedings of the 5th Nordic conference on Human-computer interaction: Building bridges* (pp. 487–490).

Kuutti, K. (2010). Where are the Ionians of user experience research? Short Paper. *Proceedings of NordiCHI 2010* (pp. 715–718). http://dx.doi.org/10.1145/1868914.1869012

Lallemand, C., Gronier, G., & Koenig, V. (2015). User experience: A concept without consensus? Exploring practitioners' perspectives through an international survey. *Computers in Human Behavior, 43*, 35–48. https://doi.org/10.1016/j.chb.2014.10.048

Lallemand, C., Koenig, V., & Gronier, G. (2013). Replicating an international survey on user experience: Challenges, successes and limitations. *Proceedings of 2013 ACM CHI—RepliCHI Workshop.*

Law, E., Hvannberg, E. T., & Hassenzahl, M. (2006, October). Theorizing, qualifying and quantifying UX. In E. Law, ET Hvannberg & M. Hassenzahl (eds.), User Experience. Towards a unified view. The 2nd COST294-MAUSE International Open Workshop, NordiCHI (Vol. 6).

Law, E. L. C., Vermeeren, A. P., Hassenzahl, M., & Blythe, M. (2007). Towards a UX manifesto. In *Proceedings of HCI 2007—The 21st British HCI Group Annual Conference* (pp. 1–2). https://doi.org/10.14236/ewic/HCI2007.95

Law, E. L. C., Roto, V., Hassenzahl, M., Vermeeren, A. P., & Kort, J. (2009, April). Understanding, scoping, and defining user experience: A survey approach. In *Proceedings of the SIGCHI Conference on Human Factors in Computing Systems* (pp. 719–728). https://doi.org/10.1145/1518701.1518813

Levi, F., Melo, P., de Lucena, U., Belleza, C., & Arcoverde, J. (2007). Evangelizing usability to 700 people: Strategies for building a user-centered organizational culture. In *Human-Computer Interaction–INTERACT 2007: 11th IFIP TC 13 International Conference, Rio de Janeiro, Brazil, September 10–14, 2007, Proceedings, Part II 11* (pp. 673–674). Springer Berlin Heidelberg.

Merholz, P. (December 13, 2007). *Peter in conversation with Don Norman about UX & innovation.* Adaptive Path. Retrieved from: http://adaptivepath.org/ideas/e000862/

Naumann, A. B., Wechsung, I., & Schleicher, R. (2009). Measurements and concepts of usability and user experience: Differences between industry and academia. In M. Kurosu (Ed.), *Human centered design* (pp. 618–626). Springer Science & Business Media. https://doi.org/10.1007/978-3-642-02806-9_72

Norman, D. A. (2010). The research-practice gap: The need for translational developers. *Interactions, 17*(4), 9–12. https://doi.org/10.1145/1806491.1806494

Norman, D. A. (2013). *The design of everyday things: The revised and expanded edition.* Perseus Books Group.

Norman, D. A., & Draper, S. (Eds.). (1986). *User centered system design: New perspectives on human-computer interaction.* Lawrence Erlbaum Associates.

Pine, B. J., & Gilmore, J. H. (1998). Welcome to the experience economy. *Harvard Business Review, 76*(4), 97–105.

Rettig, M. (1996a). Preamble. *Interactions, 3*(3), 5–8. https://doi.org/10.1145/235008.235021

Rettig, M. (1996b). Reflections on awards criteria. *Interactions, 3*(3), 64. https://doi.org/10.1145/235008.273835

Rhinefrank, J. (1995). A conversation with Don Norman. *Interactions, 2*(2), 47–55. https://doi.org/10.1145/205350.205357

Robert, J.-M., & Lesage, A. (2011). From usability to user experience with interactive systems. In G. A. Boy (Ed.), *Handbook of human-machine interaction: A human-centered design approach,* (pp. 303–320). Taylor & Francis Group.

Roto, V. (2007). User experience from a product creation perspective. In E. Law, A. Vermeeren, M. Hassenzahl, & M. Blythe (Eds.). *Towards a UX manifesto: COST294-MAUSE affiliated workshop* (pp. 31–34). Lancaster, UK.

Russell, D. M. (1998). User experience research group: Understanding the complete user interaction. *ACM SIGCHI Bulletin, 30*(2), 90–94.

Särkelä, H., Takatalo, J., Komulainen, J., Nyman, G., & Häkkinen, J. (2004, October). Attitudes to new technology and experiential dimensions of two different digital games. In *Proceedings of the third Nordic conference on human-computer interaction* (pp. 349–352). Association for Computing Machinery, New York, NY, USA

Takatalo, J., Häkkinen, J., Kaistinen, J., Komulainen, J., Särkelä, H., & Nyman, G. (2006). Adaptation into a game: Involvement and presence in four different PC-games. *Proceedings of Future-Play.*

Takatalo, J. M., & Häkkinen, J. P. (2014, October). Profiling user experience in digital games with the flow model. In *Proceedings of the 8th Nordic conference on human-computer interaction: Fun, fast, foundational* (pp. 353–356).

Tang, R., & Song, Y. (2018). Cognitive styles and eye movement patterns: An empirical investigation into user interactions with interface elements and visualisation objects of a scientific information system. *Information Research, 23*(2). https://informationr.net/ir/23-2/paper790.html

Tang, R., Lu, W., Gregg, W., Gentry, S., & Humeston, S. E. (2018). Towards a more inclusive technical website: Knowledge gaps, performance, experience, and perception differences among various user groups. *Proceedings of the Association for Information Science and Technology, 55*(1), 494–503. https://doi.org/10.1002/pra2.2018.14505501054

Tang, R., Hu, Z., Henry, N., & Thomas, A. (2021). A usability evaluation of research data management librarian academy (RDMLA): Examining the impact of learner differences in pedagogical usability. *Journal of Web Librarianship, 15*(3), 154–193.

Taylor & Francis Group. Roto, V. (2007). User experience from a product creation perspective. In E. Law, A. Vermeeren, M. Hassenzahl, & M. Blythe (Eds.), *Towards a UX manifesto: COST294-MAUSE affiliated workshop* (pp. 31–34).

Thrift, J., & Tang, R. (2018). Teaching user experience (UX) in LIS programs and iSchools in North America: Challenges and innovations. In *Proceedings of the 2018 ALISE Annual Conference* (pp. 169–174). https://digitalcommons.usf.edu/cgi/viewcontent.cgi?article=1409&context=si_facpub#page=189

Tripathi, S. (2007). User experience research at tech Mahindra. In Human-Computer Interaction–INTERACT 2007: 11th IFIP TC 13 International Conference, Rio de Janeiro, Brazil, September 10-14, 2007, Proceedings, Part II 11 (pp. 686–687). Springer Berlin Heidelberg. https://doi.org/10.1007/978-3-540-74800-7_83

Unger, R., & Chandler, C. (2012). *A project guide to UX design: For user experience designers in the field or in the making* (2nd ed.). New Riders.

UXPA. (2018, August). *2018 UX Salary Survey.* https://uxpa.org/wp-content/uploads/sites/9/2017/10/UXPA_SalarySurvey_2018v4.pdf

Worlton, W. (1971). Bulk storage requirements in large-scale scientific calculations. *IEEE Transactions on Magnetics, 7*(4), 830–833.

UX Research Models and Methods

3

Overview

In the ongoing effort to consolidate a definition of UX, scholars like Kuutti (2010) questioned whether defining UX based solely on "personal, practice-grounded opinions is really the only way to define what user experience means" (p. 716). Kuutti asks, "where are the Ionians [theorists] of UX?" (p. 716), highlighting the perceived lack of theoretical grounding in the field.

Obrist et al. (2012) echoed this concern, noting that "although there is relatively rich discussion about technological, methodological, and design issues on UX, the efforts to find and elevate the theoretical roots for UX research still fall short" (p. 1980).

Despite these critiques, various models and frameworks of UX have emerged from a range of disciplines. Obrist et al. (2012), for instance, found a special interest group session during the ACM CHI 2012 conference hat 110 respondents contributed to a total of 56 theoretical perspectives, which were mapped to seven broad categories and nine disciplines. The categories included "human/user focus," "product/artifact focus," "user/artifact/environment relations," "social nature of UX," "design focus," "frameworks involving several themes from A to E," and "even broader frameworks related to human existence." The nine disciplines the theories drew from include psychology, sociology, marketing, philosophy, communication, education, art, anthropology, and design.

Some UX models emphasize specific facets and components of user experience, while others focus on creating UX processes tailored to particular contexts, such as immersive virtual environments or news websites. Still others address the connection between product development requirements and user experience. Nevertheless, the applicability of these academic models to practitioners varies widely. While academic researchers may create frameworks that focus on theoretical dimensions of UX, practitioners in the field often develop models suited to their particular product or setting.

© The Author(s), under exclusive license to Springer Nature Switzerland AG 2025 37
R. Tang and G. Herrli, *The Domain of UX in Information Studies: Bridging Theories, Research, and Professional Practice*, Synthesis Lectures on Information Concepts, Retrieval, and Services, https://doi.org/10.1007/978-3-031-83529-2_3

Naumann et al.'s (2009) found significant differences in how academic and industry professionals approach UX research. Academic researchers tend to focus on aspects of suitability of tasks and user satisfaction, while industry professionals prioritize efficiency and aesthetics. Although both groups recognize shared criteria such as effectiveness and intuitiveness, their primary interests diverge. This divergence leads gaps in UX models: what works well in an academic contexts may not be as useful in industry, and vice versa.

The methodologies used in UX research are highly diverse, encompassing as many as 20 methods (Gray, 2016; Naumann et al., 2009; Rohrer, 2022). These methods range from qualitative, quantitative, to mixed methods design, as well as from attitudinal to behavioral techniques (Rohrer, 2022). Naumann et al. (2009) found that both academic and industry UX researchers tend to rely on a limited number of methods as their routine means of investigation, with specific methods being preferred in one setting over the other. Due to these variations in methodological preferences and familiarity, members of one community might mistrust findings produced by the other.

Academic

In the academic sphere, theoretical frameworks guiding UX research are deeply rooted in the fields of psychology, cognitive science, anthropology, sociology, cultural studies, communications, operational research, among others. Many academic researchers, particularly those from the field of Human–Computer Interaction (HCI), have inherited classical models such as GOMS (Card et al., 1983) and EPIC (Kieras & Meyer, 1997), which emphasize cognitive modeling.

During the 1990s and 2000s, alternative cognitive approaches such as distributed cognition; social approaches like Computer Supported Collaborative Work (CSCW), and imported frameworks such as activity theory became dominant in HCI. Rogers (2012) noted that in the 2000s, HCI theories underwent four directional turns: turn to design, turn to culture, turn to wild, and turn to embodiment. However, in her book entitled *HCI Theory: Classical, Modern, and Contemporary,* Rogers (2012) expressed concern that the burgeoning growth of HCI has led to a state of "flux." She indicates, "the theory that drives research is changing, a flurry of new concepts have emerged, the domains, topics and user experiences being studied have diversified" (p. 1). She also cautioned that this expansion had created "a worrying lack of direction, structure and purpose in the field," transforming what was once a focused problem space "into a more diffuse space with a less clear purpose as to what to study, what to design for and which methods to use" (p. 1).

As with HCI, UX research has seen a proliferation of conceptual approaches. As Obrist et al. (2012) observed, there diverse perspectives on the role of theories for UX design practice. The following sections will explore these various theories and methods.

Theories and Models on Experience

As discussed in Chapter "Definitions of UX", much of the conceptualization of UX is grounded in the understanding of "experience." Numerous UX frameworks and models emphasize different dimensions of this concept. Some focus on the components or building blocks that shape user experience, while others highlight the temporal aspects of UX evolves over time. These diverse approaches aim to capture the complex, multifaceted nature of user experience in different contexts.

UX Models and Frameworks

UX models that focus on the construct of experience can be categorized into two types: (1) models specifying the components of UX, and (2) those emphasizing the temporality aspect of UX. Representative models of UX components vary in their approach. Some list key variables of UX, such as Roto's (2006) illustration of UX building blocks, others outline specific components of user experience, such as Thüring and Mahlke's (2007) CUE model. Additionally, several models highlight the connection between UX and other experiences types including CX (customer experience), exemplified by Lee et al.'s (2018) structural model for experience unity.

Component models aim to illustrate the key elements ("building blocks") of UX. Roto's (2006) UX during interaction diagram, for example, identifies *User* and *System* as two primary building blocks, with their individual attributes' interaction and perceptions shaping the overall *User Experience*. The broader *Context* is seen as an overarching factor influencing UX. Similarly, the CUE model (Thüring & Mahlke, 2007) highlights *system properties*, *user characteristics*, and *task/context* as leading to *interaction characteristics,* which are connected with various UX components. The CUE section includes both *Perception of instrumental qualities* and *Perception of non-instrumental qualities,* both of which are linked to *Emotional reactions.*

These CUE components ultimately lead to an *Appraisal of the system,* influencing the overall judgments, usage behavior, and choice of alternatives, etc.

Aranyi and van Schaik (2015, 2016) adapted the CUE model to study UX in the context of news websites. They categorized variables into three groups: (a) interaction characteristics; (b) components of user experience, and (c) interaction outcomes. The components of user experience are comprised of "emotional responses (positive and negative affect), perceptions of instrumental qualities and noninstrumental qualities" (p. 2476). Aranyi and van Schaik's (2016) research also extended the model to include *trust* and *satisfaction,* finding that trust was a key predictor of behavioral intention and UX components "are strong predictors of users overall satisfaction with news sites" (p. 1563).

Meanwhile, Pucillo et al. (2014) proposed a framework connecting UX with user needs and affordances. They introduced various types of affordances, such as experience, use, effect, and manipulation affordances, which fulfill motor-goals, do-goals, and be-goals, leading to satisfaction and ultimately, pleasure. This model underscores the importance of aligning UX design with users' needs to enhance their overall experience.

Efforts to analyze and integrate various dimensions of user experience have also extended into virtual reality and augmented reality. For example, Han et al. (2018) built upon Hassenzahl's (2003) model to examine UX in the context of heritage tourism through augmented reality. They developed a specific UX model that connects product features to product character, encompassing both pragmatic and hedonic attributes. The situation forms the backdrop leading to consequences of pleasure and satisfaction.

While components-oriented frameworks are prevalent in UX research, earlier conceptualization work also emphasized the temporal aspect of UX. As early as 2006, Hassenzahl and Tractinsky (2006) pointed out UX, as a type of experience, involves "situatedness" and "temporality," extending "over time with a definitive beginning and end" (p. 94).

Meanwhile, Roto (2007) suggested that there are three phases of UX: Expected UX, UX during interaction, and overall UX. "Expected UX," formed before interaction, "plays a key role when the actual user experience takes place, as the person will evaluate the goodness of UX against the expected UX" (p. 32). Although interaction is crucial, Roto stressed that overall UX—shaped not just by interaction but also by external factors like brand image, advertising, and social influence—is critical for long-term business success. As Roto noted, "from the industry perspective, it is the long term experience that matters in business success, not a momentary emotion that might, in the end, be meaningless to the user" (p. 32).

Sutcliffe (2017) examined various UX components from the perspective of decision making when interacting with products. Sutcliffe (2017) views UX as "a process of decision making and user reaction to interactive products" (p. 106). In his model of the UX judgment process, Sutcliffe outlines three stages of UX decision-making: the first stage is "assess context," which "specifies the UX context of the product, users motivations, goals and tasks, which have been widely acknowledged to affect UX judgement" (p. 107). The second stage, "select criteria," involves criteria such as interactivity, functionality/content, brand, customization, aesthetics, and usability. The final stage is "evaluate experience," where "The criteria and context are then applied to users' perceptions with actual experience producing judgments about the quality of the UX that eventually determine product adoption" (p. 107).

Evangelos Karapanos and his coauthors have explored how UX is impacted by time (Karapanos et al., 2008, 2009, 2010). In their 2008 study, Karapanos et al. examined the initial and ongoing experiences of 10 users with an interactive TV set-top box over a four-week period, focusing on their judgments of beauty and goodness. They found that beauty judgments "were largely affected by stimulation (e.g., Novelty) during the first experiences, over time stimulation lost its power to make the product beautiful in the users' eyes" (p. 3561). In contrast, initial goodness judgments (i.e., overall satisfaction) was based on pragmatic aspects, whereas after four weeks, "identification (i.e., what the products expresses about its owner) became a dominant aspect of how good the product is" (p. 3561).

In their 2009 work, Karapanos et al. explored how UX is influenced by different phases of product adoption: *Anticipation, Orientation, Incorporation,* and *Identification.* They identified three forces driving these phases: famility, functional dependency, and emotional attachment. During the orientation stage, users' initial experiences "are pervaded by a feeling of excitement as well as frustration as we experience novel features and encounter learnability flaws" (p. 733). In the incorporation stage, "long-term usability becomes even more important … and the product's usefulness becomes the major factor impacting our overall evaluative judgments" (p. 733). As users reach the identification phase, the product becomes integrated into their lives, participating in social interactions and "communicating parts of our self-identity that serve to either differentiate us from others or connect us to others by creating a sense of community" (p. 733). At this stage, there is "an increasing emotional attachment to the product" (p. 735). Karapanos et al.'s (2009) "Framework of UX Over Time" (p. 732) effectively captures the temporality of experience, illustrating how different product qualities are valued as users transition from initial encounters to prolonged use.

Pohlmeyer et al.'s (2009) Continuous UX Model explores the temporal aspects of UX through "a series of experiential episodes spanning an entire UX lifecycle" (Kujala et al., 2013, p. 561). The ContinUE (Continuous User Experience) model divides the UX lifecycle into phases: anticipated experience (pre-use), use/experience (use), reflective experience (post-use), repetitive experience (repetitive use), retrospective experience (evaluation), and prospective experience (re-use). The model not only incorporates non-instrumental qualities but also extends "the temporal horizon of the system-user relationship" (p. 312).

Both Pohlmeyer et al.'s (2009) and Karapanos et al.'s (2009) models are referenced in Kujala et al.'s (2013) work on long-term UX. Kujala et al. (2013) emphasize "emotions and meanings" and highlight "the relationship between memories and experiences" (p. 562). Drawing from the ContinUE model and the UX overtime frameworks, their long-term UX model traces users' transition from novice to expert, focusing on learning, engagement, and relationship development, through "former usage," momentary experiences, and summarized experience. The model illustrates that UX involves "a series of episodes having consequences for the user and sometimes to society as a whole" (p. 562).

In summary, various models of UX all asserted that evaluations should extend beyond the use stage, encompassing pre-use and post-use (or overall use) experiences. These models highlight how factors such as the user, system, and context evolve over time, influencing the overall experience. By doing so, they expand our understanding of the spatio-temporal dimensions of UX (Wright et al., 2003).

Flow Theory

As a psychological model, Csikszentmihalyi's flow theory has garnered remarkable attention within the UX community, as it captures the essence of optimal user experiences. Csikszentmihalyi (1990) defines flow as "the state in which people are so involved in an activity that nothing else seems to matter; the experience itself is so enjoyable that people will do it even at great cost, for the sheer sake of doing it" (p. 4). The flow theory has been extensively tested and validated by numerous researchers (e.g., Hoffman & Novak, 1996; Koufaris, 2002; Novak et al., 2000; Guo & Poole, 2009; Alina, 2012; Barker, 2015), many of whom have investigated the antecedents and dimensions of flow.

Hoffman and Novak (1996) identified several factors that lead to flow: *control characteristics*, such as skills and challenges; *content characteristics*, such as interactivity and vividness; and *process characteristics*, such as goal-directed attributes and experiential attributes. These factors contribute to the level of involvement, focused attention and telepresence, all of which are essential for transferring an involvement into the state of flow. Guo and Poole (2009) further underscored three preconditions for flow: a clear goal, a fast and unambiguous feedback, and a perceived balance between challenge and skill.

According to Guo and Poole (2009), Csikszentmihalyi (1988) described six core dimensions of the flow experience: (1) focused concentration, (2) merging of activity and awareness, (3) a sense of being in control, (4) transformation of time, (5) a loss of self-consciousness, and (6) an autotelic experience. In comparision, Barker (2015) presented slightly different labels for some dimensions: for example, the activity is spontaneous, loss of self in the activity. Despite variations in terminology, the defining characteristics remain consistent with Csikszentmihalyi's original work.

In the UX field, flow theory is often applied in the context of immersive virtual environments. Tcha-Tokey and her colleagues (2018), through a user study involving 152 participants, collected questionnaire data to validate their model of UXIVE (User Experience in Immersive Virtual Environment). Upon analyzing the data, they revised the model based on significant results from covariance analysis. Their findings revealed that while flow is influenced by *experience, consequence,* and *skill*, "skill is the unique component influencing experience, consequence, and usability" (p. 7). The study also found that *flow, experience consequence, engagement,* and *usability* all play a role in *technology adoption*.

Researchers recognize early on that achieving flow is closely tied to factors central to usability and user experience. For instance, Skadberg and Kimmel (2004) found that contributors to flow in web browsing included "attractiveness" and "interactivity," with both "speed" and "ease of use" indirectly affecting flow. They define ease of use as "the ease with which visitors can locate the relevant information for which they are searching" (p. 414). In their analysis of 281 participants' responses regarding their browsing experience with the Great Texas Coastal Birding Trail Website, Skadberg and Kimmel (2004) discovered that speed influenced attractiveness, whereas ease of use enhanced interactivity. The authors concluded that "improving the quality of flow's contributors, attractiveness, interactivity, and their precedents, speed and ease of use, can facilitate flow experience"

(p. 418). Since these four elements—whether directly or indirectly impacting flow—are closely aligned with usability principles and UX, the connection between flow and UX becomes clear.

In Steven Pace's (2003) qualitative study on the flow experiences of 22 web users, the author found that "poor interface usability" can significantly disrupt the flow experience by introducing challenges and "divert[ing] a web user's attention from an information seeking activity, and possibly terminat[ing] a flow experience" (p. 349). Pace further emphasized that "a poorly designed interface can disrupt a flow experience by demanding an excessive amount of attention," suggesting that improving interface usability is a critical area where Web designers can reduce distractions and "maximize the opportunity to experience flow" (pp. 349–351).

User Engagement (UE) Models

The connection between flow and UX becomes even clearer through the concept of user engagement (UE). In fact, Doherty and Doherty (2018) highlight flow theory as the top theory among various models of engagement. When someone experiences flow, they are fully absorbed in the interaction, becoming completely engaged. As Sharek (2010) noted, "once engaged, a person has the opportunity to enter a flow state" (p. 26).

One specific definition of engagement links it directly with immersion and narrative flow. Douglas and Hargadon (2000) explored how interactive texts create flow, immersion, and engagement for their readers. They observed that "regardless of whether creators of interactive texts pursue chiefly flow, immersion, or engagement, readers find themselves more satisfied when the narrative brings into play their expectations and prior knowledge of other texts and schemas—even as the interactive text challenges conventions, expectations, and the range of affective pleasures alike" (pp. 158–159).

In his diagram entitled "the Engagement/Flow Cycle," Sharek (2010) illustrates a process, beginning with the user's action to "start task," followed by progress based on the presence or absence of "intrinsic or extrinsic motivation." If motivation is present, then the user initiates the engagement process, which is further fueled by the intrinsic motivation and desire. Once engaged, the participant enters a flow state, eventually exiting through disengagement.

Quesenbery (2003) suggested replacing the dimension of "satisfaction" in usability with "engagement." Quesenbery proposed five E-dimensions of usability: Effectiveness, Efficiency, Engaging, Error Tolerate, and Easy to Learn. The Engaging Dimension reflects "how pleasant, satisfying or interesting an interface is to use," and it "suggests the ways that the interface can draw someone into a site or a task," focusing on "the quality of the interaction," and the connection between the user and the product (p. 2003).

Doherty and Doherty (2018) reviewed various conceptualization of engagement, noting its links to flow, immersion, social connection and interaction, student engagement, and user involvement. In UX, user engagement (UE) is considered a "subset of UX, which focuses upon the quality of the within session interactive experience" (Hart et al., 2012, p. 1811).

Hart and colleagues measured UE through affect, aesthetics, flow/presence, usability, and preferences, while O'Brien and Toms (2008) described UE as "a quality of user experiences with technology" (p. 938), emphasizing its process from engagement to disengagement. Triggers for disengagement include usability, challenge, affect, perceived time, and interruptions, with the potential for re-engagement. O'Brien and Toms (2008), using theories like flow and play, identified attributes of UE such as "challenge, positive affect, endurability, aesthetic and sensory appeal, attention, feedback, variety/novelty, interactivity, and perceived user control" (p. 941). Lehmann et al. (2012) also viewed UE as "the quality of the user experience," focusing on the positive aspects of interaction and the user's motivation to engage with a web application (p. 164).

Similar to other definitions that interpret user engagement (UE) as a subset of UX, Oh et al. (2018) defined UE as "a form of user experience which includes both (1) a psychological state where the user appraises the quality of media and becomes absorbed in media content and (2) a behavioral experience in which the user physically interacts with the interface and also socially distributes and manages the content" (p. 742). Their research tested a four-factor conceptual model of UE—physical interaction, interface assessment, absorption, and digital outreach—using data from two experiments. The findings confirmed a continuum of UE, starting with engagement with the interface, followed by engagement with the content. The study revealed that "physical interaction and interface assessment are shown to independently predict absorption with content, which in turn predicts digital outreach" (p. 755).

Acknowledging O'Brien's work on the significance of "users' initial perception of the usability or aesthetic appeal of the interface," Oh et al. (2018) reiterated that their model "adds nuance and clarity for formally distinguishing between two different loci of engagement and the distinct user response associated with each" (p. 755). Physical interaction measures include time on hotspots and the number of actions, while interface assessment involves natural mapping, intuitiveness, and ease of use. Absorption is characterized by immersion and attention, while "digital outreach" signifies sharing and engagement activities, such as "recommend," "forward," "visit again," and 'bookmark."

It is evident that components of interface assessment and absorption are closely tied to user experience, reinforcing the idea that the essence of UX is interwoven throughout the phases of engagement.

Sutcliffe (2017) directly addressed the differences between user engagement (UE) and user experience (UX). The author explains that UE "describes how people are attracted to use interactive products ... within a session," while UX "encompasses UE but extends to the wider picture, covering why people adopt and continue to use a particular design over many sessions and even years" (p. 105). Sutcliffe (2017) further noted that while most experience ultimately revolves around utility, "to deliver utility, a product has to be easy to operate (usability) and fun (engagement)" (p. 106).

Building on this, O'Brien (2017) developed a model of UE that highlights the relationships among various factors, such as aesthetic appeal, novelty, focused attention and felt involvement, perceived usability, and endurability. O'Brien's (2017) research confirms that "aesthetic appeal and novelty predicted users focused attention and felt involvement, which predicted perceived usability. Endurability or users' overall evaluation of the experience was the outcome variable" (p. 20). This illustrates the relationship between UE, UX, and the flow experience. O'Brien (2017) reiterated that UE is "a quality of UX that is characterized by the depth of the actor's investment in the interaction; this investment may be defined temporally, emotionally and/or cognitively" (p. 22).

Academic Approaches to UX Research

In the competitive academic environment, where the "publish or perish" pressure looms large, UX researchers often focus on producing significant and quantitative-based results. Their goal is not only to demonstrate the impact of their work but also to advance scholarship and contribute to the conceptual frameworks within their field.

In Naumann et al.'s (2009) study, UX research methods in academia were compared to those used in the industry. The data revealed that questionnaires/interviews were the most common methods for academic researchers, while industry professionals favored user tests/studies.

Industry researchers also relied more on expert/heuristic evaluation, despite being familiar with questionnaires/interviews and focus groups. For academics, user tests/user studies were the second-most used method, followed by experiments, though industry researchers seldom employed experiments. As Naumann et al. (2009) noted, "'running experiments' for example is a method used primarily in academia. Paper-prototyping on the other hand seems to be a method used in industry" (p. 620). The authors also pointed out that "all groups used only few of the known and available methods" (p. 620).

In another study by Ovad and Larsen (2015), lo-fi prototyping methods, such as sketching and mock-ups, were used by 100% of the eight companies interviewed. Usability testing, including think-aloud protocols and Instant Data Analysis, was also prevalent. Notably, in 2015, workshops became the third most common method, replacing personas from 2013. An interesting finding from these studies was the new methods used by companies—AB-testing and Contextual Inquiry—The two methods were introduced through industry collaboration with academic researchers, potentially bridging the gap academic and industry UX research.

Vermeeren et al. (2010), a team of researchers from universities and industrial research labs, conducted a study on UX evaluation methods used in these settings. They gathered 96 methods from published literature, conference workshops, and SIG participants. Of these, 70% originated from academia, 18% from industry, and 12% from both. The authors noted that the lower number of industry-developed methods might be due to underreporting, as industry rarely publishes its methods. They suggested that "many industry-based methods remain unrevealed" (p. 524). Regarding data types, 39% were quantitative, 32% qualitative, and 30% mixed. UX evaluations occurred in labs 67% of the time, 52% in the field, and 40% online. Most methods (80%) involved specific user selection, and 33% involved random user selection. As to the period of the experience studied, the majority focuses on single behavioral episodes (63%), and nearly 60% conducted within one-hour test session. Fully functional products (81%) and functional prototypes (79%) dominated the methods. Methods were commonly applied to web services (81%), mobile software (77%), PC software (76%), and hardware design (66%). Most methods do not require special equipment (67%) and can often be done remotely (51%). About half need a trained researcher (49%), but 41% require minimal training. Table 3.1 provides the details of all the method dimensions and their percentages.

Upon obtaining these results, the authors grouped the methods into five categories: *specificity, utility, practicality, scoping,* and *scientific quality.* For *specificity,* only two methods were found to "explicitly studying experiences of groups of individuals" (p. 527), underscoring a need for more group-focused methods that are also feasible and "practicable." Regarding *scope,* methods for evaluating UX before product use, or "anticipated use," were identified as scarce, limiting their application in early-stage development due to concerns over "scientific quality." In terms of *practicability,* the study showed that methods are generally flexible, not constrained by the application type, development stage, or testing location. However, online evaluation methods, while promising, require improved data analysis, as current processes are "tediously time-consuming" (p. 528). For *utility,* the authors focused on expert-based "discounted methods," emphasizing the need to address their cost-effectiveness. Lastly, *scientific quality* was a concern for methods focusing on the "short term usage" experience, such as questionnaires, which were noted to have "questionable scientific quality." Even scientifically sound methods sometimes "require specific equipment, expertise and or software" (p. 528), adding further challenges.

Christian Rohrer's (2022) explored UX methods by presented a three-dimensional framework, mapping 20 popular methods across the dimensions of Attitudinal versus Behavioral; Qualitative versus Quantitative; and Context of Use (https://www.nngroup.com/articles/which-ux-research-methods/). Rohrer (2022) claims that while it is impractical to apply all methods to a single project, "nearly all projects would benefit from multiple research methods and from combining insights." Rohrer also indicates that different stages of product development call for different methods: both qualitative and quantitative methods work in the "STRATEGIZE" phase, qualitative methods are most appropriate for the "EXECUTE" stage, and quantitative methods are best utilized in the "ASSESS" stage. Nevertheless, for academic research, which often demands statistical significance, quantitative methods dominate.

Over the years, various studies have examined the research methods used in UX publications. Table 3.2 summarizes these reviews, listing methods by their frequency of use. In earlier studies, conventional methods like questionnaires, interviews, and observations

Table 3.1 UX methods as reported and examined by Vermeeren et al. (2010)

Method dimensions	Types	Percentage (%)
Originated from	Academia	70
	Industry	18
	Both settings	12
Data types	Quantitative	39
	Qualitative	32
	Mixed	30
Location of UX evaluations	Lab	67
	Field	52
	Online	40
User selection	Specific user selection	80
	Random user selection	33
	Groups	17
	Expert evaluations (no users)	14
Experience studied	Single behavioral episodes	63
	Within one-hour test session	60
	Momentary experiences	45
	Long-term experiences	36
	"Before usage" experiences	22
Product phases	Fully functional products	81
	Functional prototypes	79
	Non-functional prototypes/early designs	25
Applications	Web services	81
	Mobile software	77
	PC software	76
	Hardware design	66
Method requirements	No special equipment needed	67
	Can be done remotely	51
	Requires trained researcher	49
	Minimal training needed	41

were widely employed, along with some psychophysical measures, though less frequently. In recent years, emerging technologies such as eye-tracking, big data, and AR/VR, along with methods like user stories and user journeys, have gained prominence. Notably, no significant differences were found between methods used by researchers and industry professionals, except that students ranked biometric methods higher than industry practitioners (Lanius et al., 2021, p. 361). Several of these methods are further detailed in the "Applied UX" section.

Applied

Some Popular Techniques (e.g. Sharon, 2016)

An applied mindset begins with the concrete, and then moves to abstractions. In this section on applied methods, we'll begin with a concrete overview, explaining how applied practitioners select a research method. We will then examine how this process may differ from an academic approach.

Tomer Sharon (2016) wrote a book based on conversations with 200 applied practitioners, including startup founders, product managers, and venture capitalists. In these conversations, Sharon found that applied practitioners had eight main questions about their users: "What do people need? Who are the users? How do people currently solve a problem? What is the user's workflow? What do people want? How do people find stuff? Can people use the product? Which design generates better results?" (xxi). He structured his book around these questions, with each chapter suggesting one to three research methods to address a question, such as experience sampling to answer "What do people need?" or user interviews to answer "Who are the users?" Sharon's book format shows how applied practitioners approach research. They begin with a question about users, which they want to answer, and then they pick a research method based on that question.

Expanding on Sharon's structure, a choosable-path-adventure-style method for an applied practitioner who is choosing a research method might look something like the following:

(The initial steps labeled with S lead to common questions labeled with Q, which point to common methods labeled with M.)

Steps to Determine the Common Question
S1: Does the feature you're interested in already exist?

- No. (Or ideally no. There are questions we ask before building.)
  - Go to *Step S2: Do you know what you want to build?*
- Not yet, but we're working on it. (These are the questions we can ask while building.)
  - Go to *Step S6: What do you need to learn before developing something?*
- Yes. We want to improve something that's live on our site. (These are questions that can only be answered on the live site.)
  - Go to *Step S7: What do you want to know about this live page?*

S2: Do you know what you want to build?

Table 3.2 Review articles on UX methods used in UX research publications

Authors	Study sample	UX methods
Bargas-Avila and Hornbæk (2011)	51 publications between 2005 and 2009	• Questionnaires • Interviews (semi-structured) • User observation (live) • Video recordings • Focus groups • Interviews (open) • Diaries • Probes • Collage or drawings • Photographs • Body movements • Psychophysiological measures • Other methods
Maia and Furtado (2016)	25 publications from 2010 to 2015	• Questionnaires interview • Online survey • UX-curve • Reaction card • Observation • Reports • Video recording • Eye-tracking • Face-recognition • Brain-computer interface
Ugras et al. (2016)	199 studies published 2005–2014	• Questionnaire • Usability testing • Interview • Prototyping • Focus group • Eye tracking • Card sorting • Remote usability testing • Others
Robinson et al. (2018)	431 publications from 2000–2016	• Usability • Surveys • Interviewing methods • Ethnography • Data capture • "Other" • Focus group • Physiological • Expert review • Eye-tracking • Big data • VR/AR

(continued)

Table 3.2   (continued)

Authors	Study sample	UX methods
Tuena et al. (2020)	Systematic review of 25 publications on usability issues of virtual reality in older people	• System usability scale (SUS) • TAM-based questionnaires • UX questionnaires • UCD-based questionnaire • Flow of experience scale • Other usability questionnaires • Adherence or motivation to training • Questionnaires • Cybersickness • Assessment • Video analysis • Think aloud technique • Heuristic • Evaluation or cognitive walkthrough • Focus group • Semi-structured or structured usability postexperience interviews
Young et al. (2020)	Survey conducted in 2018 to 87 UX academic library professionals	• Usability testing • User surveys • User interviews • Accessibility evaluation • Field studies • Analytics review • Requirements and constraints • Prototype testing • Card sorting • User stories • Search log analysis • Task analysis • Persona building • Competitive analysis • Usability bug review • Journey maps • Benchmark testing • Design review • FAQ review • Diary camera studies

(continued)

Table 3.2 (continued)

Authors	Study sample	UX methods
Lanius et al. (2021)	218 publications from 2016–2018	• Surveys • Usability • Interviews • Ethnography • Biometrics • Data capture • Expert review • Focus groups • Cognitive walkthrough
Cajander et al. (2022)	Interviewed 13 UX designers about Agile UX methods	• Prototyping • User testing • User journeys • Workshops

- No. I need to figure out what to build.
  - Go to Question *Q2: What do the users need?*
- Kind of? I have an idea and want to know whether to build it.
  - Go to Step *S3: Are you wondering whether people will pay for the feature?*
- Yes. I've verified that people need this feature. Now I need to know how to build it.
  - Go to Step *S4: What do you need to understand more about before you design it?*

S3. Are you wondering whether people will pay for the feature?

(Note, this question is more market research than user research and at some organizations will thus be the responsibility of a different person.)

- Yes
  - Go to Question *Q6: Will people pay for [a product]?*
- No
  - Go to Question *Q3: Will people use [a feature]? Do people want [a product]?*

S4: What do you need to understand more about before you design it?

- Who the users are (What motivates them? What are their needs?)
  - Go to Question *Q1: Who are the users?*
- How the users act
  - Go to Step *S5: What do you need to know about how the users act?*
- Where users will expect to find this feature, and what connections do they see between it and existing features?
  - Go to Question *Q7: Where should [this feature] live?*

S5: What do you need to know about how the users act?

- How they currently solve a problem (habits they have, workarounds they've developed)
  - Go to Question *Q5: How do people currently [solve a problem]?*
- Their big-picture context of need (their actions over time, their influencers)
  - Go to Question *Q4: What is the user's (big-picture) workflow?*

S6: What do you need to learn before developing something?

- Whether people will be able to use the proposed design (or what issues they encountered using an earlier version of this feature)
  - Go to Question Q9: *Can people use [this feature]?*
- Whether people will be able to find the key calls to action (or whether they understand the key messages on the page)
  - Go to Question *Q10: Will people notice [something]?*
- Whether people can navigate to (or through) this feature
  - Go to Question *Q8: Can people find [this feature]?*

S7: What do you want to know about this live page?

- Whether people use the page
  - Go to Step *S8: Can you clarify what you mean by "use the page"?*
- People's impressions of the page (or some part of it)
  - Go to Step *S9: What about people's impressions of the page?*
- How to tweak it (to optimize for something specific)
  - Go to Question *Q12: Is ___ or ___ more [adjective]?*

S8: Can you clarify what you mean by "use the page"?

- Whether they're able to use it
  - Go to Question Q9: *Can people use [this feature]?*
- Whether they actually are using it
  - Go to Question Q11: *Do people engage with [this feature]?*

S9: What about people's impressions of the page?

- Whether people understand it
  - Go to Question *Q14: Do people know [some message]?*
- How people feel about it
  - Go to Question *Q13: How do people feel about [something]?*

Common Questions:
  Q1: Who are the users?
      When to ask

At the start of a project, and again after its launch, when there is access to usage data.

Methods to use

- M2: User Interviews
- M3: Sales Safari
- (less ideal) M4: Diary Study
- (less ideal) M5: Observation Study.

Example questions

- How do pet owners view pet-related chores?
- What motivates people to buy new outerwear?
- What distinguishes first-time homebuyers from recurring buyers?

Q2: What do the users need?

When to ask

At the start of a project, and possibly again after an unsuccessful launch.

Methods to use

- M1: Experience Sampling
- M3: Sales Safari
- (less ideal) M4: Diary Study
- (less ideal) M5: Observation Study.

Example questions

- What professional development topics are engineers most likely to be interested in?
- What frustrates or confuses parents about finding daycare for their kids?

Q3: Will people use [a feature]? Do people want [a product]?

When to ask

Before deciding whether to move forward with a new project.

Methods to use

- M6: Concierge MVP
- M7: Fake Doors Experiment
- (less ideal) M8: Crowdfunding Campaign
- (unreliably) M2: User Interviews.

Example questions

- Do people want customized news about petcare?
- Would schools people pay to have a whitelist of vetted apps?

Q4: What is the user's (big-picture) workflow?
> When to ask
>> At the start of a project
> Methods to use

- M4: Diary Study
- M5: Observation Study
- (less ideal) M2: User Interviews.

Example questions

- How do job-hunters approach finding a new job?
- How do established construction workers incorporate new best practices into their daily work?

Q5: How do people currently [solve a problem]?
> When to ask
>> At the start of a project
> Methods to use

- M5: Observation Study
- (less ideal) M2: User Interviews
- (less ideal) M4: Diary Study.

Example questions

- What workarounds have online shoppers used for finding books on a particular site?
- How do pet-lovers select which toys to buy in a pet store?
- How do plumbers determine which parts to buy in order to repair a broken shower?

Q6: Will people pay for [a product]?
> When to ask
>> At the start of a project, after scoping and before development
> Methods to use
- M8: Crowdfunding Campaign
- (less ideal) M6: Concierge MVP
- (less ideal) M7: Fake Doors Experiment

- (unreliably) M2: User Interviews.

Example questions

- Will people pay for a carol-singing food delivery service?
- Will people pay for seasonally themed recipe cards?

Q7: Where should [this feature] live?
   When to ask
      Near the end of scoping, toward the beginning of design
   Methods to use
- M9: Card Sort
- (less ideal) M10: Tree Study
- (less ideal) M13: Micro-Survey (Custom Quantitative Study)
- (less ideal) M2: User Interviews.

Example questions

- Where in the main site navigation will people expect to find information about how to tap maple trees?
- Will people expect syrup to be located with cereals or baking supplies?

Q8: Can people find [this feature]?
   When to ask
   During development
   Methods to use

- M10: Tree Study.

Example questions

- Is "bottom feeders" a category that makes sense to people seeking to buy pet fish?
- Can people find "Tomato" under "Vegetables"?
- Will people select the correct top-level menu to find information about lawncare?

Q9: Can people use [this feature]?
   When to ask
    During development, and ideally again after launch
   Methods to use

- M11: Usability Study
- M5: Observation Study.

Example questions

- Can people complete the registration form?
- Can people contact customer support?
- Can people add an item to their cart and buy it?

Q10: Will people notice [something]?
        When to ask
            During development, and ideally again after launch
        Methods to use

- M12: First Clicks Study
- (less ideal) M15: On-site Click Maps
- (less ideal) M11: Usability Study.

Example questions

- Can people find the button to sign up for the newsletter?
- Do people notice the link to the exhibit about whales?

Q11: Do people engage with [this feature]?
        When to ask
            Once a feature is live on site
        Methods to use

- M15: On-site Click Maps
- M16: On-site Analytics.

Example questions

- How many of the users in the product pages are expanding the accordion sections?
- What percentage of people click on the add-to-cart button (now vs. last month)?

Q12: Is ___ or ___ more [adjective]?
        When to ask
            Once a feature is live on site
        Methods to use

- M17: On-site A/B test.

Example questions

- Are people more likely to click "donate" or "be a donor"?
- Is adding a link to a map, or a link to an article, more effective in bringing people deeper into the site?

Q13: How do people feel about [something]?
>   When to ask
>>   At any stage, depending on the question
>   Methods to use

- M13: Micro-Survey (Custom Quantitative Study)
- M14: On-site Survey.

Example questions

- Do the product reviews help people decide what to buy?
- Which parts of the page are most useful to people?

Q14: Do people know [some message]?
>   When to ask
>>   During design or later to check understanding
>   Methods to use

- M13: Micro-Survey (Custom Quantitative Study).

Example questions

- (less ideal) M14: On-site Survey
- (less ideal) M11: Usability Study.

Methods:
M1: Experience Sampling
>   What it is
>>   Experience sampling is the method of asking people the same question repeatedly in order to identify key needs or frustrations. For example, a UX researcher can send a group of teachers multiple text messages asking "what did you most recently find frustrating about teaching?" After collecting around 1000 usable responses, the researcher can then tag and cluster the responses to identify trends.
>   Questions it answers
>>   Q2: What do the users need?
>   When to use it
>>   At the start of a project, when deciding what user problems to solve.

M2: User Interviews

What it is

A user interview involves conducting conversations with a small number of target users one at a time. In each interview, researchers ask participants a semi-structured set of questions in order to learn about their needs, motivation, and workflow. After interviewing 3–7 people for 30–60 min each, a UX researcher can analyze responses to identify trends. User interviews give context to why people do things, yet do not cover what most people do. These interviews raise questions about what people do, and these questions are answered with quantitative studies and analytics.

Questions it answers

Q1: Who are the users?

(less ideally) Q3: Will people use [a feature]?

(less ideally) Q4: What is the user's workflow?

(less ideally) Q5: How do people currently solve a problem?

(less ideally) Q7: Where should [this feature] live?

When to use it

At the start of a project, when deciding what user problems to solve. During later phases to gather context in conjunction with *M11: Usability Studies*

M3: Sales Safari

What it is

A sales safari means lurking at places online where your users talk to each other. It's a way of observing them without introducing bias through observation. For example, to see what challenges book editors face, we could read the questions they ask on a forum for book editors.

Questions it answers

Q1: Who are the users?

Q2: What do the users need?

When to use it

At the start of a project, when deciding what user problems to solve.

M4: Diary Study

What it is

A diary study gets people to record their actions over time. In this type of study, participants record their evolving needs over time, or their interactions with a product or feature. A diary study is often concluded by conducting a *M2: User Interview* with each participant.

Questions it answers

Q4: What is the user's workflow?

(less ideally) Q1: Who are the users?

(less ideally) Q2: What do the users need?

(less ideally) Q5: How do people currently solve a problem?

When to use it

At the start of a project, in order to learn about the user's context of need. After the launch of a large feature to learn how people use it.

M5: Observation Study

What it is

In an observation study, the researcher watches people complete tasks in their natural context in order to observe the difficulties they encounter, and learn about environmental factors that impact how users do things. An observation study can be fairly free-form, providing data similar to a *M2: User Interview* or short term *M4: Diary Study*. An observation study can also be structured, with specific tasks as a form of in-situ *M11: Usability Study.*

Questions it answers

Q5: How do people currently solve a problem?

Q9: Can people use [this feature]?

(less ideally) Q1: Who are the users?

(less ideally) Q2: What do the users need?

(less ideally) Q4: What is the user's workflow?

When to use it

At the beginning of a redesign, to determine what problems people encounter with an existing feature. It is also used at the beginning of a new initiative, to learn how people currently solve a problem.

M6: Concierge MVP

What it is

A Concierge MVP is a type of study that consists of a minimal version of a product, which is used to determine whether people will use the product. It often means building a front-end, and then manually implementing the back-end. For example, if a researcher wondered whether landscapers would use a resource providing custom lists of gardening tools, they could create a front-end where people input their requirements. Rather than building the back-end, they would have a person manually handle the inputs and send back a response. Once they reached a threshold, say 100 requests in a day, the researcher would know there is sufficient need for the feature, and build it out in full.

Questions it answers

Q3: Will people use [a feature]?

(less ideally) Q6: Will people pay for [a product]?

When to use it

Just before development begins, when deciding whether to create a new feature.

M7: Fake Doors Experiment

What it is

A fake doors experiment involves putting a link on the site, to a feature that doesn't exist yet, in order to see whether people are interested in that feature. It's best to use with caution, because it can result in a very negative experience for the user to be intentionally led to a dead end. As such, it's best practice to provide something for people who we lead down this path, such as giving people a way to sign up to be notified when the feature will exist, and possibly also a gift card, or a link to a similar feature.

Questions it answers

Q3: Will people use [a feature]?

(less ideally) Q6: Will people pay for [a product]?

When to use it

Just before development begins, when deciding whether to create a new feature.

M8: Crowdfunding Campaign

What it is

A crowdfunding campaign involves asking people to donate money in support of a feature. What people say they'll do in a *M2: User Interview* is often different from what they actually will do. A crowdfunding campaign is a way of seeing if people will put their money where their mouths are.

Questions it answers

Q6: Will people pay for [a product]?

(less ideally) Q3: Will people use [a feature]?

When to use it

Just before development begins, when deciding whether to create a new feature.

M9: Card Sort

What it is

A card sort involves giving a group of 30 or more study participants a set of 20–30 items, and asking them to group the items in whatever way makes the most sense to them. A closed sort is a method where the researcher gives participants the categories ahead of time. This method can be used to see whether people can identify the categories the researcher expects for new items, or to tell where new items should be located. In an open sort, the researcher asks participants to name the categories they create. It can be a useful tool for coming up with names for navigational areas. A card sort is a good method of restructuring a section of a tree between *M10: Tree Studies.*

Questions it answers

Q7: Where should [this feature] live?

When to use it

When restructuring navigation. When designing a set of new features, researchers need to know if their labels are clear. When designing a major new feature, researchers also need to know other navigation items people will associate it with.

M10: Tree Study

What it is

A tree study means giving people specific tasks and a navigational tree and, afterwards, seeing where they would look to solve the tasks. Seeing where people have the most difficulty helps to focus on parts of the tree to change. That part can be studied in a card sort, to create a new hierarchy for a follow-up tree study.

Questions it answers

Q8: Can people find [this feature]?

(less ideally) Q7: Where should [this feature] live?

When to use it

When restructuring navigation. When designing a set of new features and wondering if their labels are clear.

M11: Usability Study

What it is

A usability study means sitting a participant in front of a prototype, or a live interface, and asking him/her to complete tasks. A total of 3–5 participants is usually enough to uncover any major usability issues, although in some cases up to 8 participants may be necessary. Ideally, most tasks should be realistic tasks that the researcher knows people want to accomplish (as determined in *M1: Experience Sampling, M2: User Interviews, M3: Sales Safaris, M4: Diary Studies,* or *M5: Observation Studies*). In addition to studying user goals, some tasks can also be structured to validate usability of business goals. A usability study usually concludes with a *M2: User Interview* to obtain qualitative feedback on specific questions the researcher has about the design. A usability study is the qualitative form of a *M12: First Clicks Study.*

Questions it answers

Q9: Can people use [a feature]?

(less ideally because of the small number of data points) Q10: Will people notice [something]?

When to use it

Throughout design (and after launch if there can be further iterations).

M12: First Clicks Study

What it is

A first clicks study means giving a group of 30 or more participants an assortment of 5–10 tasks, and seeing where they would click first on a static mockup to complete those tasks. Seeing where people would click first can help to tweak the visual hierarchy of a page and can identify where labeling/messaging is unclear. A first clicks study is the quantitative form of a *M11: Usability Study.* It is like a small-scale *M15: On-site Click Map,* yet it gives context as to why people are clicking where they click.

Questions it answers

Q10: Will people notice [something]?

When to use it

During design, or before the redesign, of a live feature.

M13: Micro-Survey (Custom Quantitative Study)

What it is

A custom quantitative study is a method of finding out what people are thinking in a particular context. It's a good way of figuring out whether people understand a particular feature or label. For example, a researcher might put together a three-question survey to determine whether people understand the meaning of an on-site rating scale, and distribute it to 30 people who aren't current users of the site. Or the researcher might put together a four-question survey to see how current users feel about the site getting involved in political advocacy, and distribute it to an email list of site users. See also: *M14: On-site Survey.*

> Questions it answers
>> Q13: How do people feel about [something]?
>> Q14: Do people know [some message]?
>> (less ideally) Q7: Where should [this feature] live?
> When to use it
>> At any point to answer a narrowly scoped question that has come up.

M14: On-site Survey
> What it is
>> An on-site survey is a way of determining how people perceive something about a particular page, or about the site as a whole. For example, a researcher could use an on-site survey to benchmark whether review pages help people make decisions. On-site surveys should be used sparingly, in order to avoid annoying regular users with repeated pop-ups. See also: *M13: Micro-Survey.*
> Questions it answers
>> Q13: How do people feel about [something]?
>> (less ideally) Q14: Do people know [some message]?
> When to use it
>> Whenever there are questions about features on the live site.

M15: On-site Click Maps
> What it is
>> An on-site click map shows where people click on an existing page (or set of pages). It can also show how far down people scroll on the particular page. Additionally, it reveals which interactive elements people are (or are not) engaging with. It can also show which parts of the page people expect to be interactive, and which are currently static.
> Questions it answers
>> Q10: Will people notice [something]?
>> Q11: Do people engage with [a feature]?
>> (less ideally) Q9: Can people use [a feature]?
> When to use it
>> After launch to verify that a page is being used. Before a redesign to see what part of an existing page should be preserved or changed.

M16: On-site Analytics
> What it is
>> On-site analytics provide usage statistics about pages, as well as statistics about individual elements of pages. The most useful metrics are ratios. For example, the

total number of users of reviews will only ever go up over time. The ratio of users who came to product pages from list pages to all users of product pages can indicate how well the list is driving traffic. Similarly, the number of pageviews will only ever increase. The ratio of entrances to pageviews will indicate whether traffic is largely coming from inside or outside the site. These findings may increase or decrease over time.

Questions it answers

Q11: Do people engage with [a feature]?

When to use it

After making any changes to the live site, check how these changes impact engagement metrics (such as average time on page, average session duration, exit rate, and pages per session).

M17: On-site A/B test

What it is

A/B test is a generic term that refers to any testing of multiple variations. An on-site A/B test tweaks some elements of an existing page (or pages) to see how the change(s) will impact engagement with a page element. For example, a researcher can change the text on a button to see if a different label makes people more likely to click it.

Questions it answers

Q12: Is ___ or ___ more [adjective]?

When to use it

After a site launch, to make tweaks to a page. If during the design process, a researcher is unsure which of two things would be better to put on a page, they can add them both and use A/B testing to decide which is best.

Examining this structure indicates several things about how user research is viewed by applied practitioners:

1. Questions matter only insofar as they reveal something actionable about the users or product. Within the applied context, questions are not generally abstracted out to a level where they might apply to other situations, but instead are used to inform product decisions within a limited scope.
2. The methods fit within a product life cycle, and are thus time-bound by their need to influence the product within a set time. Rather than waiting to collect sufficient data for published results, as in an academic context, applied research is acted upon as soon as there is enough data to suggest a direction. Frequently, statistical significance is cast by the wayside, and a method only needs to point in the right direction for a product change. Continued measurement after making the change determines the perceived validity of results far more than statistical significance does.
3. Methods are only useful insofar as they answer a key question the researcher posed. Rather than following a formal structure, the method is fairly nebulously defined and can be changed to fit the question the researcher is trying to answer. A single research session can be in the form of a mash-up of usability studies, user interviews, and

moderated card sorts in order to to get feedback on a variety of questions impacting product development.

4. The methods fit throughout the product life cycle, at both the beginning and end of the process. The timing of when a method is used impacts the questions it's used for. A click map at the start of a process may be used to get a general sense of how people are using a current page, which will inform the hierarchy of the redesign. A click map created after a product launch may instead be used to determine whether people are using specific new features of the page.

5. Usability is essential; it is addressed by many of the above applied methods. Things like emotion and satisfaction are less prominent; frequently the amount of usage or engagement metrics (such as time on page, bounce rate, exit rate, and pages per session) are used in lieu of more formal methods of measuring emotion.

These trends are supported by an analysis of job postings by Garreta-Domingo and González Mosquera (2018) for UX Research positions, which identified trends in the marketplace. The researchers explain that the traditional role of a "Usability Engineer," which focuses on validating a product at the end of the development lifecycle, is increasingly being replaced by the role of a "UX Researcher," which studies user needs across all phases of product development. They found that UX researchers are expected to engage with projects throughout their entire development. The authors note that "the concept of mixed methods is also becoming widespread in research." As a result, UX researchers are expected to utilize data from both qualitative and quantitative methods. This data is gathered through studies using both methods or, alternatively, by close collaboration with colleagues who use qualitative and quantitative methods.

Lack of Formal Structure (Creating Ad Hoc Study Methods to Suit the Local Need)

Applied methods do not follow a rigidly formal structure. They can be adapted on an ad-hoc basis in order to address particular research questions. For example, software built to run card sorts for establishing a navigational hierarchy may instead be used to prioritize certain features. Instead of running card sorts, researchers may adapt the software's original function and instruct participants to drag items into categories labeled "matters to me" or "does not matter to me." A/B-testing software, originally designed to find statistical significance between two or more variations of a page, may be used as a quick way of inserting a link into a page to see how many clicks it gets. Both of the preceding examples show the efficiency and effectiveness of applying ad hoc study methods to address particular research needs.

Gaps Between Academic and Applied Models and Methods

Because of the differences in the demand between academic world and practicing realm, inevitably there are gaps between research conducted for academic purposes and for practical requirements. These issues appear in both the applied setting and in academia.

Issues with the Reliability of Methodology

In an applied setting, methodology is not often documented nor particularly rigorous.

Lack of Formal Methodology

Studies in the applied context are often structured in ways that work for researchers, with the time and resources allowed, yet these studies do not always follow specific methodologies. For example, a researcher might use a five-second impressions study with coworkers to get some preliminary feedback, then show a wireframe to a few more participants to get additional feedback. The researcher may run short on time with one of the participants, such that some questions were left unanswered. Additionally, the researcher may run into technical issues with another testing session and, as a result, the timing metrics for that particular section are invalid. Ultimately, the researcher digs into the success of the feature on the live site, using a mix of remote unmoderated usability study and click maps. The above example highlights the unreliability of research methodologies currently used in the applied context.

Lack of Statistical Significance

Studies may be cut off when a decision is needed or after a set number of participants have participated rather than once there have been enough participants to find a statistically significant effect. In applied contexts, practical significance matters more than statistical significance.

Lack of Documentation

Findings from all of these methods together may be rolled together in living documentation, which are divided among a combination of product specs, wireframes, and the live site features. Even in cases where findings are consistently documented in one place, that document likely doesn't contain information about the cultural and contextual attributes outlined above because such things are already clear to the people using the documentation.

Issues with the Applicability of Academic Approaches and Findings

A number of primary challenges that emerge when applying academic research methodologies to the professional practice of user experience (UX) includes the issue of sample size, the interpretation of the meaning of the results, the use of formal or informal methods, and the inclusion of a control group. In sections follow, we discuss the challenges in more detail.

Sample Size

In terms of the issue of sample size, in academia, researchers often struggle to secure a large enough sample to meet the statistical power required for publication in prestigious journals. Without sufficient sample sizes, the reliability and generalizability of the findings are called into question, limiting the acceptance of the research. However, in fast-paced professional settings such as UX, practitioners must often make decisions based on limited data. Time-sensitive projects cannot always wait for the ideal sample size, and decisions are frequently based on what's available rather than what is statistically ideal. This contrast between the academic insistence on statistical significance and the practical need for timely, data-driven decisions illustrates a fundamental gap between the two domains. In the applied field of UX, smaller or non-ideal sample sizes are common, but this pragmatic approach can result in research outcomes that don't align with academic standards.

The Meaning of Results

Translating findings from academic research into real-world practice is another challenge, particularly when it comes to the interpretation of results. Academic research is typically highly contextual, with results often intended to contribute to a broader body of knowledge rather than be applied directly to other settings. In contrast, UX practitioners need to apply findings in practical, often unique, contexts. A method or insight that works well in one study may not be directly transferable to a different set of users or a different product environment. This problem is particularly acute in UX, where user populations and contexts can vary widely. As a result, academic findings may not always have the immediate applicability that practitioners need, leading to a disconnect between the meaning of results in academic versus applied contexts.

Formal Versus Informal Methods

Another significant divide between academic research and UX practice lies in the distinction between formal and informal methods. Academic research is characterized by structured, rigorous methodologies designed to ensure reliability, validity, and generalizability. These methods, while valuable in producing robust findings, are often impractical for UX practitioners working in fast-paced environments. In applied settings, informal methods—such as quick usability tests, heuristic evaluations, or guerilla testing—are commonly employed. These approaches are often more flexible, faster, and tailored to the immediate needs of the project. However, because they lack the rigor of academic methods, these practices are not typically recognized in academic research. This creates a gap in how knowledge is generated and valued: methods that work well in industry may not be accepted by academia, and the formal methods championed by researchers may not be practical for professionals.

Presence or Absence of Control Groups in Applied Research

The use of control groups further underscores the challenges of bridging academic research and UX practice. In academia, the presence of control groups allows researchers to isolate variables and draw clearer, more precise conclusions. However, in applied UX research, the realities of product development cycles and user testing environments often make the inclusion of control groups impractical. Many UX studies are conducted without the rigorous controls typical of academic research, leading to findings that may be seen as less reliable by academic standards. At the same time, the absence of control groups allows UX practitioners to move quickly and gather insights that, while imperfect, are still actionable and valuable. This difference highlights the tension between the need for academic rigor and the need for fast, flexible approaches in professional practice.

Overall, the differences in how sample sizes are determined, how results are interpreted, the methodologies used, and the presence of control groups all contribute to the gap between academic research and applied UX practice. Bridging this gap requires recognizing the value and limitations of both academic rigor and the practical realities of professional UX work.

Summary

In this chapter, we explore the complexities of defining and structuring user experience (UX) research, emphasizing both theoretical and methodological aspects. Scholars such as Kuutti (2010) and Obrist et al. (2012) raise concerns over the lack of theoretical foundations in UX, with Obrist's findings revealing diverse perspectives from psychology, sociology, marketing, and design. Various UX models highlight different aspects like product development, immersive environments, user satisfaction, flow experience, and user engagement, reflecting the multifaceted nature of the field.

The chapter contrasts the theory-driven, structured methods used in academia with the practical, time-sensitive approaches in industry. Academics tend to use experimental methods and cognitive models, while industry professionals favor usability tests and heuristic evaluations. This creates gaps in how research is conducted and interpreted between the two settings.

Academic researchers emphasize statistical significance and theoretical advancements, whereas practitioners focus on actionable insights using informal methods like paper prototyping and quick usability tests. Rohrer's (2022) framework illustrates how different methods fit within various stages of product development, with less emphasis on statistical rigor in industry.

The chapter concludes by discussing the gaps between academic and applied UX research, particularly regarding sample size, result interpretation, formal versus informal methods, and the absence of control groups, underscoring the challenges of applying academic insights to real-world practice.

References

Alina, L. (2012). Developing and proposing a conceptual model of the flow experience during online information search. *Annals of Faculty of Economics, 1,* 1154–1160.

Aranyi, G., & van Schaik, P. (2015). Modeling user-experience with news websites. *Journal of the Association for Information Science and Technology, 66*(12), 2471–2493. https://doi.org/10.1002/asi.23348

Aranyi, G., & van Schaik, P. (2016). Testing a model of user-experience with news websites. *Journal of the Association for Information Science and Technology., 67*(7), 1555–1575. https://doi.org/10.1002/asi.23462

Bargas-Avila, J. A., & Hornbæk, K. (2011). Old wine in new bottles or novel challenges: A critical analysis of empirical studies of user experience. In *Proceedings of the SIGCHI Conference on Human Factors in Computing Systems* (pp. 93–102). ACM.

Barker, V. (2015). Investigating antecedents to the experience of flow and reported learning among social networking site users. *Journal of Broadcasting & Electronic Media, 59*(4), 679–697.

Cajander, Å., Larusdottir, M., & Geiser, J. L. (2022). UX professionals' learning and usage of UX methods in agile. *Information and Software Technology, 151,* 107005. https://doi.org/10.1016/j.infsof.2022.107005

Card, S. K., Newell, A., & Moran, T. P. (1983). *The psychology of human-computer interaction.* L. Erlbaum Associates.

Csikszentmihalyi, M. (1988). The flow experience and its significance for human psychology. In M. Csikszentmihalyi & I. S. Csikszentmihalyi (Eds.), *Optimal experience: Psychological studies of flow in consciousness* (pp. 15–35). Cambridge University Press.

Csikszentmihalyi, M. (1990). *Flow: The psychology of optimal experience.* Harper & Row.

Doherty, K., & Doherty, G. (2018). Engagement in HCI: Conception, theory and measurement. *ACM Computing Surveys, 51*(5), 1–39. https://doi.org/10.1145/3234149

Douglas, Y., & Hargadon, A. (2000). The pleasure principle: Immersion, engagement, flow. In *Proceedings of the Eleventh ACM on Hypertext and Hypermedia* (pp. 153–160). https://doi.org/10.1145/336296.336354

Garreta-Domingo, M., & González Mosquera, A. (2018). The evolution of UX research: A job posting analysis. *User Experience Magazine, 18*(3). Retrieved from http://uxpamagazine.org/the-evolution-of-ux-research/

Gray, C. M. (2016). "It's more of a mindset than a method": UX practitioners' conception of design methods. In *Proceedings of the 2016 CHI Conference on Human Factors in Computing Systems* (pp. 4044–4055). https://doi.org/10.1145/2858036.2858410

Guo, Y. M., & Poole, M. S. (2009). Antecedents of flow in online shopping: A test of alternative models. *Information Systems Journal, 19*(4), 369–390. https://doi.org/10.1111/j.1365-2575.2007.00292.x

Han, D., Dieck, M. C., & Jung, T. (2018). User experience model for augmented reality applications in urban heritage tourism. *Journal of Heritage Tourism, 13*(1), 46–61.

Hart, J., Sutcliffe, A., & De Angeli, A. (2012). Using affect to evaluate user engagement [Extended abstract]. In *CHI '12 Extended Abstracts on Human Factors in Computing Systems* (pp. 1811–1834). https://doi.org/10.1145/2212776.2223714

Hassenzahl, M. (2003). The thing and I: Understanding the relationship between user and product. In M. A. Blythe, K. Overbeeke, A. F. Monk, & P. C. Wright (Eds.), *Funology* (pp. 31–42). Kluwer Academic Publishers.

Hassenzahl, M., & Tractinsky, N. (2006). User experience—A research agenda. *Behaviour & Information Technology, 25*(2), 91–97. https://doi.org/10.1080/01449290500330331

Hoffman, D. L., & Novak, T. P. (1996). Marketing in hypermedia computer-mediated environments: Conceptual foundations. *Journal of Marketing, 60*(3), 50–68. https://doi.org/10.2307/1251841

Karapanos, E., Hassenzahl, M., & Martens, J.-B. (2008). User experience over time. In *CHI 2008 Proceedings* (pp. 3561–3566).

Karapanos, E., Zimmerman, J., Forlizzi, J., & Martens, J.-B. (2009). User experience over time: An initial framework. In *Proceedings of the 27th International Conference on Human Factors in Computing Systems* (pp. 729–738). https://doi.org/10.1145/1518701.1518814

Karapanos, E., Martens, J.-B., & Hassenzahl, M. (2010). On the retrospective assessment of users' experiences over time: Memory or actuality? In *Proceedings of the 28th International Conference on Human Factors in Computing Systems* (pp. 4075–4080). https://doi.org/10.1145/1753846.1754105

Kieras, D. E., & Meyer, D. E. (1997). An overview of the EPIC architecture for cognition and performance with application to human-computer interaction. *Human-Computer Interaction, 12*(4), 391–438. https://doi.org/10.1207/s15327051hci1204_4

Koufaris, M. (2002). Applying the technology acceptance model and flow theory to online consumer behavior. *Information System Research, 13*(2), 205–223. http://www.jstor.org/stable/23011056

Kujala, S., Vogel, M., Pohlmeyer, A. E., & Obrist, M. (2013). Lost in time: The meaning of temporal aspects in user experience. In *CHI '13 Extended Abstracts on Human Factors in Computer Systems* (pp. 559–564). https://doi.org/10.1145/2468356.2468455

Kuutti, K. (2010). Where are the Ionians of user experience research? Short Paper. In *Proceedings of NordiCHI 2010* (pp. 715–718). https://doi.org/10.1145/1868914.1869012

Lanius, C., Weber, R., & Robinson, J. (2021). User experience methods in research and practice. *Journal of Technical Writing and Communication, 51*(4), 350–379.

Lee, H.-J., Lee, K. K.-H., & Choi, J. (2018). A structural model for unity of experience: Connecting user experience, customer experience, and brand experience. *Journal of Usability Studies, 14*(1), 8–34. https://uxpajournal.org/wp-content/uploads/sites/7/pdf/JUS_Lee_Nov2018.pdf

Lehmann, J., Lalmas, M., Yom-Tov, E., & Dupret, G. (2012). Models of user engagement. International conference on User Modeling. Adaptation, and Personalization. 164–175.

Maia, C. L. B., & Furtado, E. S. (2016). A systematic review about user experience evaluation. In *Design, User Experience, and Usability: Design Thinking and Methods: 5th International Conference, DUXU 2016, Held as Part of HCI International 2016*, Toronto, Canada, July 17–22, 2016, Proceedings, Part I 5 (pp. 445–455). Springer International Publishing.

Naumann, A. B., Wechsung, I., & Schleicher, R. (2009). Measurements and concepts of usability and user experience: Differences between industry and academia. In M. Kurosu (Ed.), *Human centered design* (pp. 618–626). Springer Science & Business Media. https://doi.org/10.1007/978-3-642-02806-9_72

Novak, T. P., Hoffman, D. L., & Yung, Y.-F. (2000). Measuring the customer experience in online environments: A structural modeling approach. *Marketing Science, 19*(1), 22–42. https://www.jstor.org/stable/193257

O'Brien, H. L., & Toms, E. G. (2008). What is user engagement? A conceptual framework for defining user engagement with technology. *Journal of the American Society for Information Science and Technology, 59*(6), 938–955. https://doi.org/10.1002/asi.20801

O'Brien, H. (2017). Theoretical perspectives on user engagement. In H. O'Brien & P. Cairns (Eds.), *Why engagement matters: Cross-disciplinary perspectives of user engagement in digital media* (pp. 1–26).

Obrist, M., Roto, V., Vermeeren, A., Vaananen-Vainio-Mattila, K., Law, E. L-C., & Kuutti, K. (2012). In search of theoretical foundations for UX research and practice. In *Proceedings of CHI 2012* (pp. 1979–1984). https://doi.org/10.1145/2212776.2223739

Oh, J., Bellur, S., & Sundar, S. S. (2018). Clicking, assessing, immersing, and sharing: An empirical model of user engagement with interactive media. *Communication Research, 45*(5), 737–763.

Ovad, T., & Larsen, L. B. (2015). The prevalence of UX design in agile development processes in industry. In *Proceedings of 2015 IEEE Agile Conference* (pp. 40–49).

Pace, S. (2003). A grounded theory of the flow experiences of Web users. *International Journal of Human-Computer Studies, 60*(3), 327–363. https://doi.org/10.1016/j.ijhcs.2003.08.005

Pohlmeyer, A. E., Hecht, M., & Blessing, L. (2009). User experience lifecycle model ContinUE [continuous user experience]. *Der Mensch im Mittepunkt technischer Systeme. Fortschritt-Berichte VDI Reihe, 22,* 314–317.

Pucillo, F., Cascini, G., di Milano, P., di Meccanica, D., & La Masa, V. G. (2014). A framework for user experience, needs and affordances. *Design Studies, 35,* 160–179.

Quesenbery, W. (2003). Dimensions of usability: Defining the conversation, driving the process. In *Proceedings of the UPA 2003 Conference*, pp. 1–8.

Robinson, J., Lanius, C., & Weber, R. (2018). The past, present, and future of UX empirical research. *Communication Design Quarterly Review, 5*(3), 10–23.

Rogers, Y. R. (2012). *HCI theory: Classical, modern, and contemporary*. Morgan & Claypool.

Rohrer, C. (2022, July 17). *When to use which user-experience research methods*. Nielsen Norman Group. https://www.nngroup.com/articles/which-ux-research-methods/

Roto, V. (2006, October). User experience building blocks. In *The 2nd cost294-mause international open workshop* (Vol. 14, No. 1).

Roto, V. (2007) User experience from product creation perspective. In *Proceedings of Workshop on "Towards a UX Manifesto"*, 31–34.

Sharek, D. J. (2010). *Investigating real-time predictors of engagement: Implications for adaptive video games and online training*. Dissertation from North Carolina State University.

Sharon, T. (2016). *Validating product ideas: Through lean user research*. Rosenfeld Media.

Skadberg, Y. X., & Kimmel, J. R. (2004). Visitors' flow experience while browsing a Web site: Its measurement, contributing factors and consequences. *Computers in Human Behavior, 20*(3), 403–422. https://doi.org/10.1016/S0747-5632(03)00050-5

Sutcliffe, A. (2017). Designing for user experience and engagement. In H. O'Brien & P. Cairns (Eds.), *Why engagement matters: Cross-disciplinary perspectives of user engagement in digital media* (pp. 105–126). Springer.

Tcha-Tokey, K., Christmann, O., Loup-escande, E., Loup, G., & Richir, S. (2018). Towards a model of user experience in immersive virtual environments. *Advances in Human–Computer Interaction*, pp. 1–10. https://doi.org/10.1155/2018/78277286

Thüring, M., & Mahlke, S. (2007). Usability, aesthetics and emotions in human-technology interaction. *International Journal of Psychology, 42*(4), 253–264. https://doi.org/10.1080/002075907 01396674

Tuena, C., Pedroli, E., Trimarchi, P. D., Gallucci, A., Chiappini, M., Goulene, K., & Stramba-Badiale, M. (2020). Usability issues of clinical and research applications of virtual reality in older people: A systematic review. *Frontiers in Human Neuroscience, 14*, 1–19.

Ugras, T., Gülseçen, S., Çubukçu, C., İli Erdoğmuş, İ., Gashi, V., & Bedir, M. (2016). Research trends in web site usability: A systematic review. In *Design, User Experience, and Usability: Design Thinking and Methods: 5th International Conference,* DUXU 2016, Held as Part of HCI International 2016, Toronto, Canada, July 17–22, 2016, Proceedings, Part I 5 (pp. 517–528). Springer International Publishing.

Vermeeren, A. P. O. S., Law, E. L, Roto, V., Obrist, M., Hoonhout, J., & Vaananen-Vainio-Mattila, K. (2010). User experience evaluation methods: Current state and development needs. In *Proceedings of NordiCHI 2010* (pp. 521–530).

Wright, P. C., McCarthy, J. C., & Meekison, L. (2003). Making sense of experience. In M. Blythe, A. Monk, C. Overbeeke, & P. C. Wright (Eds.), *Funology: From usability to user enjoyment* (pp. 43–53). Kluwer.

Young, S. W., Chao, Z., & Chandler, A. (2020). User experience methods and maturity in academic libraries. *Information Technology and Libraries, 39*(1), 1–31.

Sharing and Communicating UX Research

4

Overview

In this chapter, we examine how UX empirical discoveries are disseminated, shared, and presented in both academic and industry settings. By comparing scholarly discourse with professional reports, we identify gaps between the two domains and offer recommendations for bridging this divide.

Academic

In an academic setting, the dissemination of UX research findings is primarily achieved through the publication of research papers or reports in scholarly journals, or by presenting conference papers at international, national, or regional conferences. Both avenues typically involve a peer-review process, but they differ in terms of timeline and revision demands. Journal submissions often require multiple rounds of rigorous review and revision, which can extend the time to publication. In contrast, conference papers usually adhere to strict submission deadlines and undergo blind review by two to three reviewers. The review process for conferences tends to be quicker, often concluding after one round, and authors usually have an opportunity to revise their papers before submitting the final version for presentation.

Beyond papers and conference presentations, UX researchers may also share their work by writing or editing books or contributing chapters on these topics. These books can range in size and format, from concise professional booklets to comprehensive textbooks, depending on the intended audience and scope of the subject matter. Whether for academic instruction or professional development, books provide another vital platform for disseminating UX research in academic settings.

© The Author(s), under exclusive license to Springer Nature Switzerland AG 2025
R. Tang and G. Herrli, *The Domain of UX in Information Studies: Bridging Theories, Research, and Professional Practice*, Synthesis Lectures on Information Concepts, Retrieval, and Services, https://doi.org/10.1007/978-3-031-83529-2_4

Characteristics of Scholarly Discourse of UX

The following sections outline the characteristics of UX scholarly discourse by examining various metrics and data patterns. These include the number of UX-related publications over the years, the volume of cited references on the topic, the keywords used in these publications and cited references, and a comparison of the keywords used in UX research across two key disciplines: Computer Science—CS (the leading subject area for both publications and citations) and Social Sciences (SS). The data used in this section was obtained from searching SCOPUS in September 2024 and analyzing relevant outputs and results concerning the number of publications or cited references, subject areas, keywords, and more.

Number of Publications Over the Years

In September 2024, the first author conducted a search on SCOPUS for publications with titles or abstracts containing the terms "user experience" or "usability." The data revealed a clear trend of exponential growth in publications related to these topics, especially after 2000, with notable increases over the past decade. This surge reflects the rising integration of technology into daily life and the growing emphasis on enhancing user experiences in software, digital systems, and various interactive platforms.

Specifically, the breakdown of publications by five-year intervals can be seen in Table 4.1.

An analysis of the growth trend reveals that there is a rapid growth in recent years (2020–2024), with the number of publications surging to close to 62,000 publications in 2020–2024, comparing to less than 42,000 in 2015–2019, representing a 49% increase. Note that the most significant publication volume was in 2023, close to 15,000 publications, suggesting an increasing interest in "user experience" and "usability" studies. From 2010–2019, there is a steady growth. Specifically, from 2010 to 2014, there were close to 27,000 publications, increasing to close to 42,000 in 2015–2019, a growth of about 54%. This period marks a significant expansion in these research areas.

In early 2000s, there is a moderate growth, with the total number of publications between 2000 and 2009 grew steadily, from 5476 in 2000–2004 to 15,752 in 2005–2009, indicating a 188% increase. This period likely reflects the growing adoption of digital systems and the emergence of human–computer interaction as a field. Prior to 2000, however, there were minimal UX publications. The number of publications was sparse, with only 2256 in 1995–1999 and even fewer before then. The field was in its infancy, with usability and user experience being niche topics.

In last 30 years, the UX publications has an exponential growth. From just 42 publications in 1965–1969, the field has grown exponentially to 61,957 by 2020–2024. This explosive growth reflects the increasing importance of user-centered design in technology development.

Table 4.1 Number of publications with title words or abstracts containing "user experience" or "usability"

Year range	Year	Number of publications	Total
2020–2024	2024	11,945	61,957
	2023	14,816	
	2022	12,775	
	2021	11,659	
	2020	10,762	
2015–2019	2019	10,269	41,496
	2018	8823	
	2017	8163	
	2016	7440	
	2015	6801	
2010–2014	2014	6301	26,973
	2013	6141	
	2012	5301	
	2011	4841	
	2010	4389	
2005–2009	2009	4080	15,752
	2008	3555	
	2007	3155	
	2006	2765	
	2005	2197	
2000–2004	2004	1617	5,476
	2003	1275	
	2002	958	
	2001	828	
	2000	798	
1995–1999	1999	532	2,256
	1998	505	
	1997	487	
	1996	378	
	1995	354	
1990–1994	1994	264	1,064
	1993	234	
	1992	203	

(continued)

Table 4.1 (continued)

Year range	Year	Number of publications	Total
	1991	185	
	1990	178	
1985–1989	1989	142	587
	1988	107	
	1987	94	
	1986	131	
	1985	113	
1980–1984	1984	104	418
	1983	91	
	1982	79	
	1981	82	
	1980	62	
1975–1979	1979	66	271
	1978	58	
	1977	53	
	1976	51	
	1975	43	
1970–1974	1974	36	162
	1973	42	
	1972	29	
	1971	21	
	1970	34	
1965–1969	1969	17	42
	1968	9	
	1967	4	
	1966	5	
	1965	7	
1960–1964	1964	4	12
	1963	6	
	1962	1	
	1961	1	
1955–1959	1959	3	15
	1958	3	

(continued)

Table 4.1 (continued)

Year range	Year	Number of publications	Total
	1957	2	
	1956	3	
	1955	4	
1950–1954	1954	3	11
	1953	6	
	1952	2	
1945–1949	1949	2	5
	1948	1	
	1945	2	
1940–1944	1941	1	1
1935–1939	1939	1	2
	1936	1	
1930–1934	1931	1	1
1920–1924	1921	1	1

Number of Citations Over the Years

In September 2024, the first author conducted a search on SCOPUS for cited references with title containing the terms "user experience" or "usability." The data shows a significant and exponential increase in the number of UX cited references in academic research, particularly after 2010. This growth coincides with the increasing recognition of user-centered design in the development of digital technologies and services. The period from 2020 to 2024 represents the most substantial growth, with more than 146,000 cited references related to UX, indicating that these topics are now critical in research and technology development. Specifically, the breakdown of cited references by five-year intervals can be seen in Table 4.2.

An analysis of the growth trend reveals that in recent years (2020–2024), there is an exponential growth in UX related cited references in recent years, with the number of cited references has exploded in this period, with more than 146,000 UX citations. There has been a rapid increase, particularly in 2023 with close to 35,000 cited references, which is a new high, and 2024 also showed a large number with more than 28,000 citations. In 2015–2019, there is a substantial growth. From close to 77,000 citations, this five-year period represents a significant growth over the previous one (2010–2014), with a roughly 108% increase in citations from 2010–2014 to 2015–2019, reflecting increased academic interest and research activity in these topics.

The period from 2010 to 2014 saw a steady growth, with a 40% growth compared to the previous five years (2005–2009), with 36,864 citations, reflecting the continuing rise in research involving user experience and usability studies. From 2005–2009, there

Table 4.2 Number of cited references with title words containing "user experience" or "usability"

Year range	Year	Number of cited references	Total
2020–2024	2025	93	**146,435**
	2024	28,398	
	2023	34,844	
	2022	31,099	
	2021	28,282	
	2020	23,719	
2015–2019	2019	20,679	**76,748**
	2018	16,985	
	2017	14,969	
	2016	12,975	
	2015	11,140	
2010–2014	2014	9653	**36,864**
	2013	8284	
	2012	7040	
	2011	6300	
	2010	5587	
2005–2009	2009	5032	**17,416**
	2008	4023	
	2007	3481	
	2006	2688	
	2005	2192	
2000–2004	2004	1590	**4,864**
	2003	1175	
	2002	872	
	2001	677	
	2000	550	
1995–1999	1999	402	**1,535**
	1998	373	
	1997	297	
	1996	236	
	1995	227	
1990–1994	1994	164	**698**
	1993	159	

(continued)

Table 4.2 (continued)

Year range	Year	Number of cited references	Total
	1992	139	
	1991	120	
	1990	116	
1985–1989	1989	75	**305**
	1988	64	
	1987	65	
	1986	54	
	1985	47	
1980–1984	1984	32	**161**
	1983	48	
	1982	26	
	1981	30	
	1980	25	
1975–1979	1979	19	**71**
	1978	11	
	1977	18	
	1976	11	
	1975	12	
1970–1974	1974	4	**19**
	1973	8	
	1972	3	
	1971	2	
	1970	2	

is a gradual growth, and citations doubled in this period compared to 2000–2004, from 4864 to 17,416, indicating more established research interest in these topics as technology became more user-centric.

In the early 2000s, there is an initial uptick, with citations increased significantly from 1,535 in 1995–1999 to 4864, indicating that the fields of "usability" and "user experience" began to gain traction during the early 2000s. Before 2000, UX-related citations were relatively low. The 1990s (1535 total) showed some early interest, while the 1980s (305 total) and earlier periods had very minimal activity in these research areas.

Overall, the growth pattern is exponential. From the early 2000s, the number of UX related cited references has increased dramatically, reflecting the mainstreaming of user experience and usability studies across multiple disciplines.

Subject Areas of Publications and Cited References

When looking at the distribution of the subject areas of publications or cited references, the patterns are very revealing. In both publications and cited references, CS dominates with the highest numbers. CS not only contributes a large volume of UX research but also plays a critical role in informing research across various fields. While UX research may originate primarily from fields like CS and Engineering, it has far-reaching influence, particularly in Medicine, Social Sciences, and Business. UX-related research has a substantial citation footprint, reflecting its importance in interdisciplinary applications.

Subject areas reflected in UX publications

Table 4.3 presents the distribution of subject areas of publications where titles or abstracts contain the terms "user experience" or "usability," underscoring the interdisciplinary nature of UX research. The data reveal that CS leads with close to 90,000 publications, highlighting its pivotal role in the development of digital systems, software, and user interfaces. Engineering ranks second with close to 50,000 publications, demonstrating the significance of usability in hardware design, industrial systems, and technology-driven solutions. Medicine and Social Sciences also feature prominently, reflecting the growing focus on user experience in healthcare technologies and the social impacts of technological advancements.

Other key fields include Mathematics, Physics and Astronomy, and Decision Sciences, indicating that UX research is grounded in both scientific and analytical frameworks. The influence of UX is also notable in Business, Management, and Accounting and Economics, Econometrics, and Finance, signaling its importance in market-driven solutions and financial services. Furthermore, the fields of Health Professions, Biochemistry, Genetics, and Molecular Biology, and Psychology emphasize the relevance of UX in human-centered and biological sciences.

The distribution of publications across such diverse subject areas demonstrates the expansive reach of user experience research, extending beyond the traditional technology sectors into healthcare, business, education, and scientific disciplines.

Subject areas reflected in UX cited references

Table 4.4 illustrates the distribution of subject areas where cited reference titles contain the terms "user experience" or "usability," showcasing the extensive reach of UX research across multiple disciplines. As with publications, CS dominates, with more than 125,000 cited references, demonstrating its critical role in driving advancements in UX research and technology-focused innovations. Engineering (58,255) and Medicine (57,890) follow closely, highlighting the importance of usability not only in technological developments but also in healthcare systems.

Social Sciences also holds a significant share, with more than 55,000 cited references, reflecting the growing attention to user experience from a societal and human behavior perspective. Noteworthy contributions come from fields like Mathematics, Biochemistry,

Table 4.3 Subject area distribution for publications with title or abstract containing "user experience" or "usability"

Subject area	Number of publications
CS	89,624
Engineering	48,644
Medicine	23,836
Social Sciences	22,490
Mathematics	22,358
Physics and Astronomy	8778
Decision Sciences	7388
Materials Science	7046
Business, Management and Accounting	6262
Environmental Science	5163
Biochemistry, Genetics and Molecular Biology	4672
Health Professions	4648
Energy	4036
Arts and Humanities	3893
Earth and Planetary Sciences	3627
Psychology	3393
Chemistry	3208
Chemical Engineering	3135
Agricultural and Biological Sciences	2539
Nursing	2482
Neuroscience	1955
Economics, Econometrics and Finance	1454
Multidisciplinary	1376
Pharmacology, Toxicology and Pharmaceutics	1018
Immunology and Microbiology	668
Dentistry	209
Veterinary	195
Undefined	14

Genetics, and Molecular Biology, and Agricultural and Biological Sciences, underscoring the integration of UX principles in scientific and biological research.

Additionally, the business sector has a substantial presence, with close to 20,000 cited references in Business, Management, and Accounting, indicating the relevance of

Table 4.4 Subject area distribution for cited references with title containing "user experience" or "usability"

Subject area	Number of cited references
CS	125,101
Engineering	58,255
Medicine	57,890
Social Sciences	55,002
Mathematics	25,125
Biochemistry, Genetics and Molecular Biology	21,599
Agricultural and Biological Sciences	20,934
Business, Management and Accounting	19,550
Environmental Science	13,773
Decision Sciences	11,805
Psychology	11,646
Arts and Humanities	10,512
Health Professions	9567
Physics and Astronomy	8825
Immunology and Microbiology	8112
Materials Science	7648
Nursing	7244
Energy	6135
Neuroscience	5743
Earth and Planetary Sciences	5535
Chemistry	5418
Economics, Econometrics and Finance	5365
Chemical Engineering	5148
Multidisciplinary	5039
Pharmacology, Toxicology and Pharmaceutics	2845
Veterinary	1074
Dentistry	532
Undefined	15

UX in market-driven solutions, management strategies, and business applications. Disciplines like Decision Sciences and Psychology emphasize the importance of human factors and decision-making in shaping user experiences. Fields with a health focus, including Health Professions, Nursing, and Neuroscience, reflect the growing role of UX in medical technologies and healthcare delivery systems.

Overall, the distribution of cited references across these diverse subject areas highlights the interdisciplinary influence of UX research, extending from technology and engineering to the social sciences, health, and business domains.

Comparative analysis of subject areas reflected by UX publications and cited references
The comparison of subject area distribution between publications and cited references on "user experience" or "usability" reveals interesting patterns as follows:

- CS dominates both publications and cited references, emphasizing its foundational role in UX research and influence across disciplines.
- Engineering and Medicine both rank high in publications and citations, highlighting UX's significant impact in healthcare and technology.
- Social Sciences, Psychology, Mathematics, and Decision Sciences contribute fewer publications but are heavily cited, underscoring UX's relevance in human behavior, social impact, and quantitative research.
- Business, Biomedical, and Environmental Sciences show fewer publications but high citation counts, reflecting UX's importance in business strategy, scientific tools, and technology.
- Arts, Humanities, and Health-Related Fields UX research is less frequently published but widely cited, particularly in healthcare, emphasizing UX's growing role in these areas.

In summary, UX research, heavily rooted in CS and Engineering, significantly influences Medicine, Social Sciences, and Business, with its interdisciplinary applications evident across many fields. Citations often outpace publications, reflecting UX's wide-reaching impact.

Keywords of UX Publications and Cited References in CS and Social Science

Among the UX publications and cited references, CS dominates both publications and cited references, where social science's outputs are also high. A comparison between keywords used in CS and Social Science might reveal good insights. The following discussion addresses the patterns based on the common keywords shared between the two disciplines and unique keywords from each of the two disciplines. There are a total of 163 keywords from CS UX publications, whereas Social Science UX publications contain 127 keywords.

Common Keywords in UX publications in CS and Social Science
There are 69 common keywords between CS and SS UX publications. These include terms such as "User Experience", "Usability", "Human, "User Interfaces", "Ergonomics",

"E-learning", "Virtual Reality", "Design", "Augmented Reality", and "Artificial Intelligence." Table 4.5 shows some of the top common keywords. The comparison of keyword frequency rankings between CS and SS in UX publications reveals distinct differences in focus and priorities across the two disciplines:

- "User Experience" ranks 1st in CS and 2nd in SS. While both fields recognize the importance of UX, CS places it at the forefront, likely due to its role in software

Table 4.5 Common keywords of UX publications from CS and social science

Common keywords	CS frequency	CS frequency rank	SS frequency	SS frequency rank
User experience	14,436	1	2026	2
User interfaces	10,926	2	1448	4
Human computer interaction	10,060	3	1063	10
Usability	7225	4	2119	1
Usability engineering	6574	5	1317	5
Virtual reality	6052	6	950	13
Design	3784	8	711	18
Websites	3290	9	599	22
Augmented reality	3105	10	495	26
Artificial intelligence	3093	11	375	35
Learning systems	3031	12	621	20
E-learning	2974	13	1,060	11
Visualization	2677	15	327	38
Students	2577	19	1178	8
Human	2506	20	1819	3
Ergonomics	1103	72	1107	9
Surveys	2877	14	668	19
Adult	676	121	844	15
Education	1492	52	819	16
Teaching	1056	74	580	23
Software engineering	2098	27	213	77
Machine learning	2021	29	226	68
Mobile computing	1830	34	212	80

development, interface design, and user interactions with technology. In SS, although also highly ranked, the slightly lower position may reflect a broader focus on the social implications and human aspects of user experience.

- "Usability" ranks 4th in CS but 1st in SS, indicating a much stronger emphasis in SS on ensuring that systems, tools, and products are accessible and functional for users.
- "Human–Computer Interaction (HCI)" ranks 3rd in CS and 10th in SS. This may show that CS focuses heavily on the technical and design aspects of how humans interact with computers, whereas SS approaches it from a broader, less frequent perspective, likely emphasizing the sociocultural and behavioral aspects of HCI.
- "User Interfaces" ranks 2nd in CS and 4th in SS, highlighting its importance in both disciplines. However, the higher rank in CS suggests a stronger technical focus on creating and optimizing interfaces, while SS may analyze their social impact and usability for different demographic groups.
- "Virtual Reality (VR)" ranks 6th in CS and 13th in SS, reflecting a more technology-driven interest in CS where VR plays a key role in advancing interactive experiences. In SS, the interest in VR is likely connected to studying its effects on behavior, education, or social interaction.
- "Augmented Reality (AR)" ranks 10th in CS but 26th in SS, showing that CS is significantly more focused on the development and technical aspects of AR. In contrast, SS has a much lower frequency, likely investigating AR's social or educational implications but with less technical focus.
- "Design" ranks 8th in CS and 18th in SS. The higher ranking in CS indicates a strong focus on the technical and practical aspects of design within digital systems, whereas SS may take a broader view, examining the societal implications of design practices but with less frequency.
- "Artificial Intelligence (AI)" ranks 11th in CS and 35th in SS, which reflects CS's strong focus on AI as a driving force behind technological innovations in UX. SS, by contrast, explores AI in fewer instances, likely in the context of its impact on users, ethics, and social systems.
- "E-learning" ranks 13th in CS and 11th in SS, showing its relevance in both fields but with slightly more emphasis on SS, where the focus might be on the effectiveness of digital learning tools, user engagement, and education technologies. In CS, it is more related to the development and optimization of e-learning systems.
- "Human" ranks 20th in CS but 3rd in SS, which underscores Social Science's focus on human factors, behaviors, and experiences in UX studies. In CS, "Human" ranks lower, suggesting a more system-centric approach with human factors considered but not as central to the research focus.

In summary, Computer Science prioritizes technical and development-related keywords such as "HCI," "User Interfaces," "VR," "AR," and "AI," reflecting its focus on building and optimizing interactive systems and technologies. Social Science, on the other hand, places greater emphasis on "Usability," "Human Factors," and the social dimensions of UX, demonstrating its broader, user-centered approach to understanding how technology impacts behavior and society. Both fields recognize the importance of "User Experience" but approach it from different angles: technical in CS and human-centered in SS.

Unique Keywords in UX publications in CS and Social Science
Ninty-four keywords appear only in CS UX publications. Some examples include: "Deep Learning," "Internet of Things," "Mobile Computing," "Software Engineering," "Recommender Systems," "Big Data," "Wearable Technology," "Robotics," "Data Privacy," "Speech Recognition," and "Neural Networks." On the other side, 58 keywords appear only in SS UX publications. Some examples include: "Psychology," "Sustainability," "Qualitative Research," "COVID-19," "Focus Groups," "Perceived Usability," "Mental Health," "Older Adults," "Satisfaction," "Social Media," "Sustainable Development," and "Young Adult."

   Overall, CS UX publications focus more on technical aspects such as "Software Engineering," "Deep Learning," and "Data Privacy," reflecting the field's emphasis on developing and refining the technology behind user experiences. SS UX publications prioritize human-centered issues, focusing on "Psychology," "Sustainability," "Qualitative Research," "Mental Health," and demographic-specific studies such as "Older Adults" and "Young Adult." This highlights the field's focus on user impact, behavior, and societal implications of UX design.

Common Keywords in UX cited references in CS and Social Science
Table 4.6 reflects some of the top common keywords in UX cited references in CS and Social Science. The comparison between **CS** and **SS** UX cited references reveals patterns where some keywords have similar rankings, while others differ significantly.

Keywords with Similar Rankings
There are two to three keywords that have similar frequency ranking. "E-learning" ranks 7th in both CS and SS, showing a shared focus on digital learning technologies, though CS emphasizes system development while SS explores educational outcomes. "Decision Making" ranks almost equally (19th in CS and 18th in SS), reflecting its importance in both fields, with CS focusing on decision support systems and SS on cognitive processes. Lastly, "Surveys" ranks 14th in CS and 19th in SS, indicating both fields use surveys rather frequently in usability testing and system evaluations.

Keywords with Dissimilar Rankings
There are about 15 keywords in cited references that have rather dissimilar frequency ranking. For instance, "Human–Computer Interaction (HCI)" is 1st in CS but 11th in SS, illustrating CS's technical focus on interactions with computers, while SS places less emphasis on this but considers broader human aspects. "User Interfaces" ranks 2nd in

Table 4.6 Top common keywords in UX cited references in CS and in social science

Common keywords	CS frequency	CS frequency rank	SS frequency	SS frequency rank
Human computer interaction	14,415	1	1863	11
User interfaces	11,258	2	1756	13
Virtual reality	8595	3	2036	9
User experience	7286	4	1381	15
Usability	6256	5	1887	10
Design	5763	6	1376	16
E-learning	5309	7	2309	7
Usability engineering	5291	8	1082	23
Human	5009	11	5102	1
Students	5015	10	2505	4
Female	1539	68	2409	5
Male	1481	72	2400	6
Adult	1367	80	2252	8
Education	2752	27	1827	12
Ergonomics	1282	91	1449	14
Teaching	1965	46	1315	17
Decision making	3569	19	1270	18
Surveys	4773	14	1220	19

CS but 13th in SS, also highlighting its technical importance in CS, with SS focusing on social impact rather than development. Meanwhile, "Virtual Reality" ranks 3rd in CS and 9th in SS, indicating CS's stronger focus on VR development, while SS explores its social implications. "User Experience" is 4th in CS and 15th in SS, reflecting CS's emphasis on technical UX design, whereas SS adopts a broader, human-centered approach. "Usability" ranks 5th in CS but 10th in SS, important in both fields, though framed technically in CS and more user-centric in SS. "Design" ranks 6th in CS and 16th in SS, where CS focuses on system design and SS on social dimensions of design.

"Usability Engineering" is 8th in CS but 23rd in SS, indicating greater technical emphasis in CS.

On the other hand, "Human" ranks 1st in SS but 11th in CS, showing SS's central focus on human behavior, while CS concentrates more on systems. "Students" is 4th in SS and 10th in CS, with SS giving more importance to education and learning behaviors. Gender-related terms ("Female" and "Male") rank far higher in SS (5th and 6th) compared to CS

(68th and 72nd), reflecting SS's emphasis on gender studies. Similarly, "Adult" ranks 8th in SS and 80th in CS, showing a stronger focus on adult populations in SS. "Education" is 12th in SS but 27th in CS, where SS prioritizes educational outcomes, while CS is more focused on developing tools for education. "Ergonomics" ranks 14th in SS but 91st in CS, with SS placing more value on human-centered design principles. "Teaching" ranks 17th in SS and 46th in CS, with SS focusing more on pedagogical approaches.

Overall, UX cited references in CS focuses more on technical and system-related keywords, including "User Interfaces," "Usability Engineering," and "HCI," while UX cited references in SS emphasizes human-centered and societal keywords, such as "Human," "Gender," and "Students." The overlap in keywords like "E-learning" and "Decision Making" may reveal shared interests but with different disciplinary approaches.

Journal Papers, Conference Presentations, and Book Publishing

In an academic setting, the primary avenue for disseminating UX research is through journal articles and conference proceedings. For faculty members, the number of publications is a primary indicator of research productivity, serving as evidence of academic achievement and playing a crucial role in tenure and promotion evaluations. According to the publication data from SCOPUS, from 1980 to 2025, UX publications predominantly appeared in journal articles and conference papers, with a modest yet increasing presence in books and book chapters, although they still represent a small share.

As shown in Table 4.7, journal articles and conference papers have consistently represented a significant majority of UX publications, comprising between 91.25 and 95.63% of total publications over the decades. The peak was observed in the 1980s and 1990s, where they accounted for 95.62–95.63%. In recent decades, this figure has slightly declined to around 91.25–91.70%. Conversely, the proportion of books and book chapters has gradually increased, from just 0.30% in the 1980s and 1990s to 2.85–3.78% in the 2000s and 2010s.

Table 4.7 Number and percentage of UX publications in journal articles/conference papers format and in books/book chapters format

Year range	Total publications	Journal articles/conference papers (%)	Books or book chapters (%)
2020–2025	62,003	56,747 (91.52%)	1892 (3.05%)
2010–2019	68,469	62,784 (91.70%)	2588 (3.78%)
2000–2009	21,228	19,371 (91.25%)	605 (2.85%)
1990–1999	3320	3175 (95.63%)	10 (0.30%)
1980–1989	1005	961 (95.62%)	3 (0.30%)

Peer-Review Cycle and Length

As noted earlier, over 90% of UX scholarly output consists of journal articles or conference papers. While the number of books and book chapters has been increasing, they represent less than 5% of the overall production. Journal articles and conference papers typically undergo a peer review process.

The peer review process is vital for ensuring the scientific rigor of published research. Depending on the journal or conference, review models may vary and can include blind (single-blind or double-blind) or open formats. In a *single-blind* review, the reviewers know the authors' identities, while the authors do not know who the reviewers are. In a *double-blind* review, neither party knows each other's identity. An *open* review model allows both authors and reviewers to know each other's identities. The peer review process for journal manuscripts typically involves the steps outlined in Table 4.8. Conference paper reviews may involve fewer steps, with acceptance decisions often made in a single round.

The duration of the peer review process varies based on publication type and discipline. For journal articles, the entire process—from initial review to final publication—typically takes 3–6 months, though it can extend to a year or longer in some cases. In contrast, conference paper reviews generally follow a more compressed timeline, often lasting about

Table 4.8 Steps in a typical peer review process

Step	Description
1. Submission	Authors submit their manuscripts to a journal or conference, following specific formatting guidelines
2. Initial editorial review	The editorial team conducts a preliminary assessment to determine if the submission fits the scope of the journal or conference and meets basic quality standards
3. Reviewer selection	Editors select experts to evaluate the work, ensuring no conflicts of interest
4. Conducting the review	Reviewers assess the manuscript's originality, significance, methodology, and clarity
5. Reviewer feedback	Reviewers provide evaluations and suggestions for improvement to the editor
6. Editorial decision	Editors make a decision: "Accept," "Minor Revisions," "Major Revisions," or "Reject."
7. Revision	Authors revise the manuscript and resubmit, possibly undergoing additional reviews
8. Final decision	Editors make a final decision based on the revised manuscript and additional comments
9. Publication	Accepted articles are prepared for publication, including editing and formatting

1 month. Authors are usually notified of decisions simultaneously at the end of the review period, with some conferences choosing not to allow revisions due to tight schedules. Papers are either accepted or rejected based on the initial submission. Generally speaking, the peer review process for conference papers is faster due to the fixed deadlines associated with conference schedules, whereas journal articles may undergo multiple rounds of review and revision, leading to longer durations.

Academic Presentations

When a UX conference paper or research article has been accepted for journal publication, authors typically give a presentation at a conference or being invited to present at a research forum. A typical conference presentation generally lasts 15–20 min, followed by a Q&A session. Some webinar session may last for an hour, with the presentation usually begins with a brief introduction, outlining the research question and its significance. Following this, the presenter summarizes the methodology, key findings, and implications of the research, often using visual aids such as slides or posters to enhance understanding. Presenters typically aim for clarity and engagement, encouraging audience participation during the Q&A segment. Effective academic presentations also address potential limitations and future research directions, providing a comprehensive overview of the study's contribution to the UX field.

Presentation to Stakeholders

Other than academic audiences, UX researchers are often asked to present their findings to various stakeholders. According to Kuniavsky (2003), when presenting usability or UX findings to different audiences, different approaches should be considered: "Different audiences require different approaches when presented information. The exact needs of each audience are going to be unique, but there are some generalizations that can be made about certain specific groups." (P. 496). Based on Kuniavsky's tips, Table 4.9 outlines different practitioner groups' characteristics and the corresponding presentation foci.

Applied

While the academic publication cycle culminates in publication, written documentation is of lower significance in applied UX. Applied practitioners write only as much as necessary in order to recall decisions made at a later time, and to communicate key concepts to people not involved in preliminary meetings. Applied practitioners treat written documentation as something that evolves over time, rather than something to be written once, finalized, and distributed. In an applied context, written documentation is often used as a "visual record" that preserves memory. As Hoffman (2018) writes: "get those notes in front of everyone's eyes at the same time, in real time while the discussion is happening. While people are engaged in talking[...], the scribe creates a visual record of only the

Table 4.9 UX presentation to different groups

Stakeholder group	Characteristics	Presentation tips
Software engineers	• Problem solvers • Familiar with scientific methodologies • Skeptical of non-statistical methods	• Solution-focused: highlight specific solutions • Balance context: discuss broader issues and actionable changes • Concrete facts: provide solid data • Prioritization: explain how research prioritizes user needs
Visual designers	• Creative problem solvers • Transitioning from identity to interaction design • More concerned with face validity	• Identity versus functionality: distinguish product identity from functionality • Problem prioritization: emphasize prioritizing issues • Immediate sense: ensure recommendations are intuitive
Marketing researchers	• Focused on consumer desires • Utilize quantitative tools	• Reasons behind choices: explain motivations behind behaviors • Cohesive model: present a unified model of user behavior • Predictive insights: show how research forecasts behavior
Upper management	• Concerned with long-term strategy • Involved in product and business details • Request metrics to evaluate success	• Interests and agenda: tailor to their specific interests • Critical metrics: focus on relevant metrics for decision-making • Clear findings: present key findings clearly • Time for questions: allow time for engagement

Source Kuniavsky (2003, pp. 496–501)

main ideas, the conflicts, and the decisions on the wall. It needs to be large enough so that anyone in the room can read it from wherever they are sitting. That way, when the scribe captures something incorrectly, someone in the meeting can speak up and provide a correction" (p. 33). When written documentation is created by a group in order to preserve group memory, it doesn't have the same cachet that it does in academic circles. In these settings, written documentation is treated as a missive, rather than a tangible certificate of accomplishments. As such, as ideas change, the meanings of written documentation change naturally over time.

Because the end goal of the applied practitioner is to create a viable product, any documentation that does not directly contribute to the success of that product is superfluous. This means that while applied practitioners run competitive analyses, which are similar to academic literature reviews, these analyses are not explicitly necessary in every case.

Competitive analyses do not need to follow a formal structure. These analyses link to resources for the sole purpose of providing concrete examples, rather than \citing sources. Because applied documentation is almost never published for outside audiences, the formalized tracking of sources is unnecessary. Similarly, while applied practitioners create deliverables that gives an overview of study findings, these deliverables differ from academic study findings. These deliverables evolve over time, and only document findings that have a direct impact on the development of the product. The end goal of such a deliverable is to ensure team members internalize the research findings, so that they can use them to develop the product. As such, bringing those team members into the initial collection of data is a way of ensuring that the research is acted upon, while simultaneously minimizing the need for written documentation.

Living Documentation

As design has grown in applied practice, outside of the rigid academic context, the need for documentation has shifted. As Gothelf and Seiden (2013) write: "lean UX refocuses the design process away from the documents the team is creating to the outcomes the team is achieving.[…] Documents don't solve customer problems—good products do" (p. 12). Rather than documenting results in an academic report at the end of a process, the applied Lean process documents things along the way. Gothelf and Seiden write, "externalizing means getting your work out of your head and out of your computer and into public view. Teams use whiteboards, foam-core boards, artifact walls, printouts, and sticky notes to expose their work in progress to their teammates" (p. 10).

Bringing the Team into the Research (E.G. Buley, 2013)

In *The User Experience Team of One*, Leah Buley emphasizes that a single person with an official UX title can effectively act as part of a much larger UX team by bringing non-UX colleagues \ into UX work. Buley writes, "every day is an opportunity to invite your non-UX colleagues into the world of UX[…] treat them as partners in the ongoing project of making your products as user-friendly as possible" (p. 43). In the applied context, official areas of expertise are less important, and what becomes important instead is getting the full team involved and on the same page. In such organizations, UX practitioners are not solely experts in UX, but "facilitators and conduits of ideas held by an entire cross-functional team" in order to "create a design solution that cleverly reconciles those tensions and produces a satisfying experience for users" (p. 45).

Getting Various Groups to Internalize Research Findings

One goal of bringing non-UX colleagues into UX research is to get them to understand research findings. According to Buley, "colleagues need to better understand the value of UX. Invite them into your process and yourself into theirs[....] When conducting user research, invite your non-UX colleagues to come along" (p. 47). The second author (Graham) finds that inviting colleagues to user research sessions is a profoundly effective way to garner understanding about user research findings. When other colleagues watch the participants themselves, they empathize with those participants much more strongly than if they just heard about the participants in a report afterward. As a result, there is a need for greater empathy with the goals of the end user, which outweighs the need for documentation.

Another method for getting team members to understand research findings are collaborative design meetings. A common setting for collaborative design meetings are design studios, where user researchers present study findings to show user needs and behaviors, product managers present business goals, and cross-functional groups will sketch their ideas for an interface addressing those needs. After everyone sketches, there is a structured activity to share and critique the sketches, and then a vote on a design to move forward with. The collaborative design meeting structure ensures that all of the major stakeholders in the process are involved early on, and that the entire team internalizes research findings by drafting their own version of how those findings could impact the final product. As Gothelf and Seiden (2013) note, "collaborative design is an approach that allows a team to create product concepts together. It helps teams build a shared understanding of the design problem and solution.[...] Collaborative design is still a designer-led activity. It's the designer's responsibility not only to call these meetings but to facilitate them as well" (pp. 34–35).

Gaps Between Academic and Applied Approaches to Communicating UX Research

Academic Versus Applied Approaches to Documenting UX Research and Data Analysis

In academic settings, documenting UX research involves a rigorous, often lengthy peer review process. Research must be presented in detail, with every component carefully described to ensure transparency and replicability. In contrast, applied UX research focuses on succinct, actionable insights designed to meet the immediate needs of stakeholders such as designers, developers, and executives. The following highlights key differences between the academic and applied approaches.

Academic Approach

- **Thorough Documentation**: Academics prioritize comprehensive documentation, including extensive literature reviews, detailed methodologies, exhaustive data analysis, and concrete results. This ensures the research process is transparent and replicable.
- **Peer Review and Publication**: The primary goal of academic research is to publish in peer-reviewed journals or conference venues, which require rigorous and thorough documentation to establish credibility and reliability.

Applied Approach

- **Concise Reporting**: In applied settings, reporting is concise and efficient. Stakeholders expect key findings and actionable insights without the exhaustive detail typical of academic work.
- **Time Constraints**: Applied UX research often operates under tight deadlines, requiring minimal documentation that still conveys essential findings and recommendations clearly and effectively.

Generalizability Versus Transferability of Results

Academic research typically emphasizes the generalizability and theoretical implications of its findings, aiming for broad applicability across multiple contexts. In applied research, the focus is on how findings can be translated into practical solutions for specific projects or environments.

Academic Approach

- **Generalizability**: Academic research aims to produce findings that are applicable across various contexts, often relying on large, diverse sample sizes to ensure broad relevance.
- **Theoretical Contributions**: The focus is on contributing to a broader body of knowledge, often emphasizing developing theories and principles that can be applied universally.

Applied Approach

- **Transferability**: Applied research is concerned with how findings can be transferred or adapted to specific projects or contexts. Case studies are often used to provide detailed insights that can be directly applied.
- **Practical Solutions**: The emphasis is on providing context-specific solutions, rather than developing broad generalizations, ensuring that findings are relevant and actionable.

Making Concrete Interface Improvement Recommendations

In academic research, the focus is often on theoretical and long-term implications, which can make it challenging to provide specific recommendations for interface design improvements. Applied research, on the other hand, is geared toward actionable solutions that can be implemented quickly.

Academic Approach

- **Theoretical Focus**: Academic research tends to prioritize theoretical insights, which can limit its ability to make specific, concrete recommendations for user interface improvements.
- **Long-Term Impact**: Academics often focus on long-term implications, which can be less immediately actionable for practitioners working on real-world projects.

Applied Approach

- **Actionable Insights**: Applied research aims to provide clear, actionable recommendations for improving user interfaces, often tailored to the specific product or service being developed.
- **Direct Application**: The goal is to directly enhance the user experience of a particular product, making the research highly relevant and immediately useful to stakeholders.

Bridging the Gap Between Academic and Applied UX Research

Understanding these gaps allows both academic and applied researchers to better tailor their communication strategies to meet the needs of their respective audiences. This can engender more effective collaboration and ensure that UX research findings are successfully applied in real-world settings. In the next chapter, a framework is proposed to address various dimensions where the gaps may be bridged, but in terms of UX results' communication and publication, below is an example of the bridge that can built to fill the gaps.

An example of efforts to bridge these gaps is the editorial approach of *Journal of User Experience (JUX)*, which includes sections such as "Recommendations" and "Tips for Usability Practitioners." These sections are designed to make research findings more accessible and implementable for usability professionals. The journal's guidelines suggest providing three to five tips for practitioners, typically drawn from the method, findings, or conclusions, ensuring the research remains both academically rigorous and practical useful. As stated in "JUX Manuscript Guidelines" (2022), the "Tips for Usability Practitioners" section is "one of the unique features of the Journal of User Experience," and the purpose is "to extract ideas from your manuscript that usability practitioners can apply in their current work." The guidelines further clarify that the tips are aimed at "user experience practitioners who are not researchers," highlighting that "Tips are your way to emphasize a practice or finding that practitioners might be able to apply in their work" (p. 8). This deliberate strategy fosters the translation of UX research into insights that can directly benefit usability professionals.

References

Buley, L. (2013). *The user experience team of one: A research and design survival guide.* Rosenfeld Media

Gothelf, J., & Seiden, J. (2013). *Lean UX: Applying lean principles to improve user experience.* O'Reilly Media Inc.

Hoffman, K. M. (2018). *Meeting design: For managers, makers and everyone.* Rosenfeld Media.

Journal of User Experience. (2022). Preparing a JUX Manuscript: Guidelines for Authors. https://uxpajournal.org/submit/

Kuniavsky, M. (2003). *Observing the user experience: A practitioner's guide to user research.* Elsevier.

Where to Go from Here: Bridging the Gaps

5

Suggestions for Academic Researchers to Impact Applied Work

Academic researchers in UX have a unique opportunity to shape the future of user experience design by aligning their research closely with real-world applications. To maximize their practical impact, they should focus on producing actionable insights that practitioners can implement directly. This involves conducting studies that address current challenges in the field, exploring under-researched areas, and presenting findings in a manner that is accessible and relevant to UX professionals.

One effective approach to bridging this gap is to engage in collaborative research projects with industry partners or UX practitioners. Such partnerships enable researchers to test their theories and models in practical settings, resulting in more impactful research outcomes and practical solutions. Additionally, academic researchers should consider simplifying their language and emphasizing the practical implications of their work when publishing or presenting their findings. This approach will ensure that their research resonates with those working in applied contexts.

Research Partnership with UX Practitioners in the Field

Strong partnerships between academic UX researchers and UX professionals in the field are essential for advancing the discipline and bridging the gap between theory and practice. UX practitioners offer valuable insights into real-world challenges and provide a current view of industry trends, which can inspire new research questions and directions for academics. Meanwhile, academic researchers bring theoretical insights and empirical rigor, helping to validate and refine the methods used in applied settings.

These partnerships may take various forms, including:

- **Co-designing research projects**, where UX professionals present real-world challenges and academics contribute theoretical frameworks and rigorous methodologies to address these issues.
- **Establishing regular communication channels** including workshops, conferences, or online forums that facilitate knowledge-sharing, discussion of findings, and the exploration of new ideas between both groups.
- **Engaging in longitudinal studies** that track the impact of UX interventions over time, yields valuable data for academic analysis and practical application in the field.

Through these collaborative efforts, academic researchers gain access to practical data and real-world scenarios, while UX professionals benefit from evidence-based approaches to solving design problems. This symbiotic relationship ensures that research is grounded in real-world applications and that UX professionals stay informed about the latest theoretical advancements and scientific knowledge.

Ultimately, these partnerships foster the development of more robust UX methodologies, tools, and frameworks that are not only theoretically sound but also highly relevant and actionable in practice. This alignment strengthens the field, leading to more impactful research and practical, innovative solutions for UX challenges.

UX Designers Providing Real Life Scenarios for Academic Researchers to Investigate

Real-life scenarios and challenges provided by UX designers in applied settings can be a valuable source of data for academic research. These scenarios offer insight into the practical difficulties faced by UX professionals, enabling researchers to design studies that directly address these issues. By examining these real-world situations, academic researchers can generate findings that are immediately relevant and impactful for the field.

To foster this collaboration, UX practitioners should be encouraged to share their experiences and challenges with the academic community. This can be achieved through case studies, workshops, or collaborative projects. By working with practical data, researchers can develop more applicable theories and offer insights that help UX designers improve their practice and apply evidence-based solutions.

A Call to Academically Validate and Formalize Methods as They Are Used in the Field

Many UX methods in the field have been developed through practical application. While these methods may prove effective in real-world settings, they might lack the rigorous testing and formalization that academic research provides. There is a need for academia to validate and formalize these methods, ensuring they are grounded on solid theoretical frameworks and supported by empirical evidence.

This process involves not only testing the effectiveness of existing methods but also refining them to enhance their reliability and validity. By academically validating these methods, researchers can contribute to the development of standardized practices in the field, which can lead to more consistent and reliable UX outcomes. This, in turn, strengthens the credibility of UX as a discipline and ensures that practitioners are using the most effective tools available.

Suggestions for Workers in Applied Fields to Strengthen Their Academic Foundation

For UX researchers in applied settings, a strong academic foundation is essential for staying competitive in the field. Understanding the theories and research underlying UX practices enhances a UX practitioner's ability to apply these methods effectively. Practitioners are encouraged to engage with academic literature, attend conferences, and participate in professional development courses that emphasize the theoretical aspects of UX.

By building a solid academic foundation, UX professionals can make more informed decisions in their work, ultimately leading to better outcomes for users. Moreover, a deeper understanding of the academic side of UX can create opportunities for collaboration with researchers, further bridging the gap between theory and practice.

Incorporating Statistical Patterns to Represent Distribution of Results

Utilizing statistical tools such as error bars and distribution curves in UX research can strengthen the rigor and credibility of findings. Error bars indicate data variability, offering insights into the reliability of the results, while distribution curves visually depict how data points are spread across a range. These elements provide a more nuanced understanding of the data, helping both researchers and practitioners in applied settings to make better-informed decisions.

Incorporating statistical rigor into UX research and practice is essential for producing reliable, valid results. UX professionals and academic researchers may consider:

- **Including error bars or distribution curves** in reports and presentations to clearly communicate the variability and reliability of their findings.

- **Using appropriate statistical methods** to analyze UX data, ensuring conclusions are grounded in sound evidence.
- **Continuously educating themselves on statistical concepts** to better interpret research findings and apply them in their work.

By adopting these practices, UX professionals can elevate the quality of their research, foster transparency, and make decisions that are more informed by the variability and reliability of their data. This approach not only strengthens the credibility of UX findings but also helps in making informed decisions based on data.

Project Partnership with Academic Researchers

Collaborative partnerships between UX professionals in the field and academic researchers offer a powerful way to bridge the gap between theory and practice. These collaborations enable UX practitioners to tap into cutting-edge research and academic rigor, while academics gain access to real-world data and practical application opportunities. Together, they create a dynamic environment for mutual learning and innovation.

Such partnerships can be structured around specific projects, with each party contributing their unique expertise. For instance, a UX project within a company could involve academic researchers who bring theoretical insights and methodological rigor, ensuring the project not only addresses practical challenges but also contributes to the broader scholarly discourse. This collaborative approach enhances project quality by validating methodologies and ensuring results are both practical and reliable.

Key benefits of these partnerships include:

- **Bringing academic rigor** to practical projects, ensuring that research methodologies are robust and outcomes are credible.
- **Providing access to cutting-edge research** and innovative tools that may not be readily available within the industry.
- **Fostering innovation** by combining practical UX expertise with the latest theoretical frameworks, driving the development of new products, services, and processes.

Ultimately, these collaborations lead to the creation of more effective and innovative solutions, strengthening both academic research and UX practice.

Suggestions for Instructors and Curriculum Designers to Bridge Between Academic and Applied UX

Faculty members, instructors and curriculum designers play a crucial role in preparing the next generation of UX professionals. To effectively bridge the gap between academic and applied UX, educational programs must be designed to integrate both theoretical knowledge and practical skills. In the sections follow, a set of strategies are outlined to achieve this balance.

Include Both Academic Oriented Research Literature and Practical Hands-On Projects in Classes

In higher education, UX curricula must incorporate both academic research literature and practical, hands-on projects. This combination provides students with a comprehensive education, ensuring they understand foundational concepts and theories while gaining practical experience. Furthermore, UX curricula should be interdisciplinary, integrating fields such as psychology, CS, and design. This reflects the multifaceted nature of the profession, equipping students with a broad skill set that enhances their adaptability and innovation in their careers.

In a panel article resulting from a tutorial workshop during the ASIS&T 82nd Annual Meeting (2019) in Melbourne, Fei Yu et al. (2020) gathered a panel of UI/UX instructors from academia and industry to discuss "Innovative UI/UX methods for information access based on interdisciplinary approaches." The panelists emphasized the need to integrate theory with practice in UX education. Javed Mostafa, one of the panelists, stated that "Regarding the UI/UX curriculum for an iSchool, it needs to achieve a good balance between theory and practice" (p. 76). Another panelist, Qian Xu, explained how this balance is operationalized in their courses: "To discuss key theories and concepts, I pulled content from different disciplines, such as sociology, anthropology, psychology, information science, communication studies, and graphic design. I try to incorporate more hands-on activities and assignments that are closely related to the introduced theories and concepts" (p. 76). Xu also acknowledged the challenge of providing an entry-level overview of various UI/UX topics but noted that efforts have been made to include industry-relevant content: "In addition to academic articles for reading, I started to include industry reports, white papers, and even well-written blogs in writing assignments, which worked well" (p. 76).

Table 5.1   UX course final projects. *Source* Thrift and Tang (2018)

Execution final project	Team	Individual		
	72%	28%		
Final product	Research study	Design of systems	Other	
	51%	46%	3%	
System evaluated	Live system	Hypothetical system	Real client	Not specified or other
	29% (12)	26% (11)	16% (7)	29% (12)
Involve practical, hands-on activities	100%			

Real-World Term Projects in UX Courses

Research by Thrift and Tang (2018) examined 67 ALA-accredited LIS programs and nine iSchools in North America, analyzing UX-related courses and surveying faculty members who teach them. Among 39 UX courses with final projects, over half required conducting a UX research study or usability evaluation, while the rest focused on interface or website design. One-third of the final projects involved working with a live system, another third involved designing hypothetical systems, and over 15% required working with a real client. Of the 12 courses that involved a live system, most conducted user research, with some proposing redesigns. Courses using hypothetical systems focused on design-based projects, while projects with real clients were split between design and research. The findings highlight that even six years ago, 100% of the UX courses included "practical, hands-on activities" and were moving toward using real systems with real clients (Table 5.1).

Invite Speakers from Both Academic and Applied Settings to Present Challenges in Their Work

Inviting speakers from both academic and industry settings can give students a well-rounded perspective on the field of UX. These guest speakers provide valuable insights into the challenges they face in their work, the strategies they use to overcome them, and how theory and practice intersect in their day-to-day tasks.

Such presentations help students understand the real-world relevance of their academic studies, offering practical examples of how to apply their knowledge. This exposure also demystifies the field of UX, illustrating that professionals successfully navigate its complexities by integrating both academic and practical knowledge.

In a panel article discussed earlier, several panelists emphasized their connections with the industry and how they regularly invite UX professionals as guest speakers. As Fei

Yu noted, "Since most of my students have very strong career goals in terms of desiring more knowledge about UI/UX practice in the industry and internship opportunities, I have developed and maintained collaborations with a few working professionals in the industry through Carolina alumni networks. So far, I have invited five guest speakers from Nielsen Norman Group, Optum/UnitedHealth, Bank of America, Facebook, and Netflix to routinely join my classes either physically or remotely" (p. 77).

A Call for Sharing Research Findings and Best Practice Discoveries to Avoid Duplicated Work

The UX field thrives on innovation and continuous learning, yet the absence of consistent knowledge-sharing mechanisms can lead to duplicated efforts and missed opportunities for progress. To advance the discipline and optimize resources, the UX community must foster a culture of open collaboration, where research findings and best practice discoveries are shared widely and effectively.

The culture of sharing within the UX community may take following forms:

- **Open Access to Research Findings and Best Practices**: Publishing research findings and sharing best practices openly—whether through journals, conferences, or online platforms—ensures that valuable insights are accessible to everyone. This transparency prevents the duplication of efforts, enabling researchers and practitioners to build on existing knowledge rather than starting from scratch. By offering free or low-cost access to this information, the community encourages continuous learning and improvement.
- **Establishing Collaborative Networks**: Forming collaborative networks where academics and practitioners can regularly exchange insights, tools, and resources will strengthen ties between the two communities. These networks provide a platform for ongoing dialogue, where real-world challenges can inspire new research directions and academic findings can directly influence practice. Through such exchanges, the UX community becomes more cohesive, accelerating the flow of knowledge across disciplines.
- **Centralized Repositories for UX Research and Case Studies**: A centralized, shared repository of UX research, case studies, and best practices would serve as a vital resource for both academics and practitioners. This repository could house a wide range of content, from empirical studies to practical case studies, making it easier for UX professionals to access proven strategies and insights. By reducing redundant research efforts, such a platform would help fast-track innovation and ensure that both new and seasoned professionals benefit from collective knowledge.

By embracing these strategies, the UX community can foster an environment where knowledge flows freely between academia and industry, accelerating advancements while minimizing wasted efforts. In turn, this culture of sharing strengthens the field, ensuring that innovations are informed by a wide array of perspectives and that solutions are developed with a comprehensive understanding of existing knowledge.

A Framework for Bridging the Gaps

Based on the previous discussion, the authors propose a framework that addresses the critical disconnects between academic research and applied practice in User Experience (UX) by fostering collaboration, education, validation, and continuous feedback. The framework provides a structured approach that leverages theoretical insights and practical application to ensure that UX evolves as a discipline that is not only innovative but also deeply relevant to both scholars and practitioners.

The specific components, their purposes, and key elements are outlined in Table 5.2, it is also illustrated in Fig. 5.1, with the component "knowledge sharing and open access" visualized as a floating component that crosses over of the other four components. Furthermore across various components, the main goals of bridging the gaps via "infusion," "translational development," and "alliances," as the terms used by Dray (2009) and Norman (2010), are showing in the diagram. Both Dray and Norman's works were discussed in detail in Chap. 2 of this book, in the subsection "Overview" of the section "Gaps between academic and applied definitions and conceptualizations of UX."

The specification of each of the components are described as follows:

Component 1. Collaboration Between Academia and Industry
Purpose: To promote joint research and practice-based projects that address real-world challenges.

Collaboration is the cornerstone of this framework. By encouraging partnerships between academics and industry professionals, this component ensures that UX research is grounded in the realities of practice while giving practitioners access to the latest academic insights. These collaborations can take various forms:

- **Collaborative Research Initiatives**: Academics and practitioners work together from the outset to design research projects that are both theoretically robust and practically relevant.
- **Joint Research Projects and Case Studies**: Addressing actual UX problems in the industry helps apply academic insights to real-world challenges, benefiting both parties.
- **Internships, Mentorship Programs, and Partnerships**: Creating pathways for students and early-career professionals to work alongside industry experts ensures that they gain hands-on experience while still benefiting from academic mentorship.

Table 5.2 A framework for bridging the gap between academia and practice in UX

Component	Purpose	Key elements
Collaboration Between Academia and Industry	To promote joint research and practice-based projects that address real-world challenges	• Collaborative research initiatives involving academics and UX practitioners from the outset • Joint research projects and case studies that solve practical UX problems • Internships, mentorship programs, and industry partnerships
Integrated Curriculum Design	To provide students with both academic knowledge and practical skills, preparing them for the UX profession	• Curriculum blending interdisciplinary knowledge from psychology, design, and CS • Real-world, hands-on projects and case studies • Guest lectures and workshops from UX professionals and academics
Feedback Loops for Validation and Adaptation	To continuously test and refine UX methods, ensuring they are both academically valid and practically effective	• Field testing of academic findings by practitioners to validate usability and relevance • Practical insights and real-world data fed back into academic research for refinement • Continuous adaptation of research methodologies based on real-world performance

(continued)

Table 5.2 (continued)

Component	Purpose	Key elements
Ongoing Professional Development	To ensure continuous learning and mutual knowledge exchange between academia and industry	• Professional development programs designed for both UX professionals and academics • Short courses, seminars, and workshops focused on current industry trends and academic advancements • Lifelong learning opportunities for career growth
Knowledge Sharing and Open Access	To create a culture of open communication that accelerates innovation and reduces duplicated efforts	• Open-access platforms for sharing UX research, case studies, and best practices • Collaborative networks and repositories for ongoing knowledge exchange • Conferences, workshops, and online forums fostering a shared dialogue between academia and industry

Such collaborative efforts bridge the gap by blending theoretical knowledge with real-world practice, leading to innovations that are informed by both spheres.

Component 2. Integrated Curriculum Design

Purpose: To provide students with both academic knowledge and practical skills, preparing them for the UX profession.

Education plays a pivotal role in preparing future UX professionals who are well-versed in both theory and practice. This component emphasizes creating curricula that reflect the interdisciplinary nature of UX, combining knowledge from fields like psychology, design, and CS.

Key strategies include:

- **Blending Interdisciplinary Knowledge**: Courses should include theoretical foundations from multiple disciplines relevant to UX, fostering a broader understanding of human behavior, systems, and design thinking.
- **Hands-on Projects and Case Studies**: Real-world projects should be incorporated into coursework, allowing students to apply the theories they learn to practical challenges they will encounter in the field.
- **Guest Lectures and Workshops**: Inviting industry professionals and academics to share their experiences ensures students are exposed to both perspectives and better prepared for their careers.

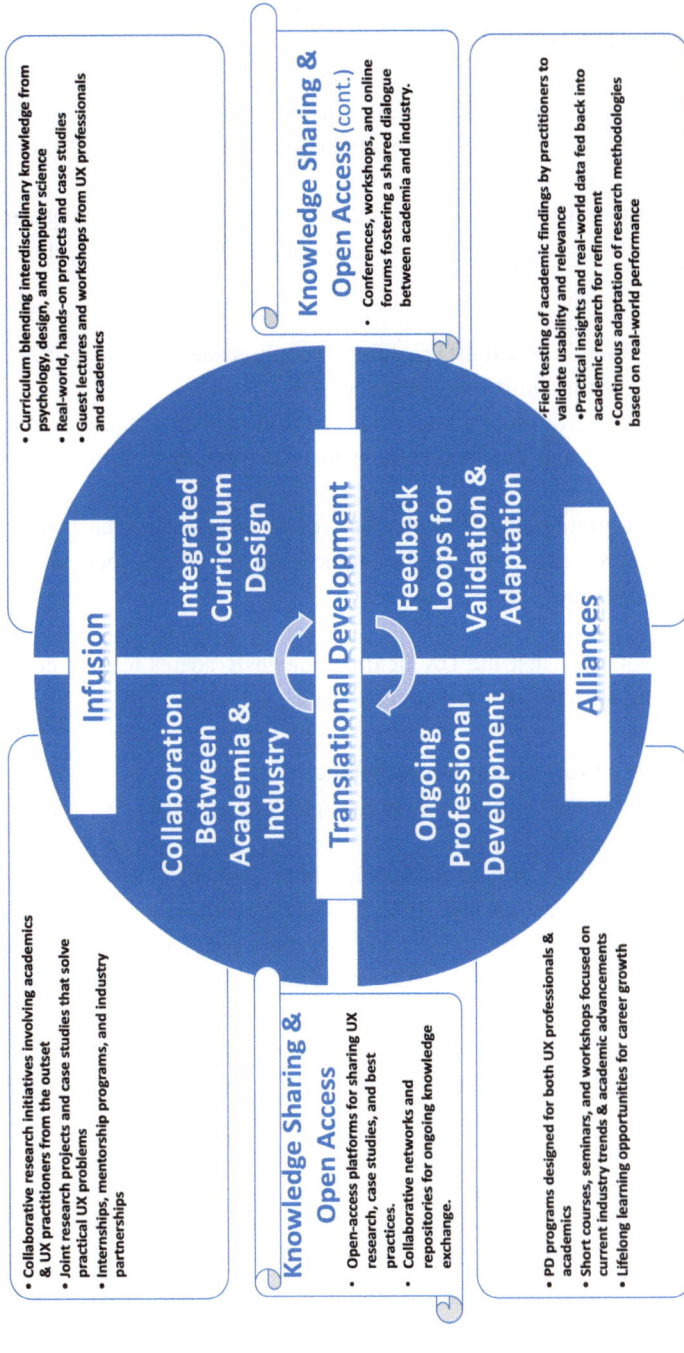

Fig. 5.1 A framework for bridging the gap between academia and practice in UX

By integrating academic knowledge with practical skills, educational institutions can bridge the gap between the classroom and the workplace, equipping students to tackle the complexities of the UX profession.

Component 3. Feedback Loops for Validation and Adaptation
Purpose: To continuously test and refine UX methods, ensuring they are both academically valid and practically effective.

This component ensures that theoretical research is tested in real-world settings, and practical insights are fed back into academic research, creating a continuous cycle of refinement and improvement.

- **Field Testing of Academic Findings**: Practitioners apply academic findings in real-world settings to validate their usability and relevance.
- **Real-World Data Integration**: Insights from practical applications are used to refine research methodologies and theories, making them more applicable to everyday UX challenges.
- **Methodological Adaptation**: As UX evolves, this component ensures that both academic and industry practices remain current and aligned, with feedback loops allowing for ongoing adjustments based on performance data.

These feedback loops ensure that the UX field continuously adapts to new insights and challenges, maintaining its relevance and effectiveness.

Component 4. Ongoing Professional Development
Purpose: To ensure continuous learning and mutual knowledge exchange between academia and industry.

Professional development is critical for keeping both academics and practitioners up-to-date with the latest advancements in UX. This component fosters lifelong learning opportunities and creates spaces for professionals to share insights, challenges, and best practices.

- **Professional Development Programs**: Short courses, seminars, and workshops tailored for both academics and practitioners ensure that both groups stay current with industry trends and academic developments.
- **Knowledge Exchange**: Mutual learning opportunities where academics and professionals collaborate on projects or attend cross-disciplinary learning sessions.
- **Lifelong Learning Opportunities**: UX is a rapidly evolving field, and practitioners need continual upskilling to stay relevant. Offering flexible learning pathways, such as certifications or refresher courses, supports career-long development.

Through continuous learning and interaction between academia and industry, both groups benefit from new perspectives, and the profession remains innovative.

Component 5. Knowledge Sharing and Open Access
Purpose: To create a culture of open communication that accelerates innovation and reduces duplicated efforts.

Finally, fostering a culture of openness is key to avoiding duplicated work and accelerating progress in UX. By promoting the sharing of research findings, case studies, and best practices, this component ensures that the field of UX evolves in a collaborative and inclusive way.

- **Open-Access Platforms**: Researchers and practitioners should be able to easily access UX-related research and findings through open-access publications, online repositories, and other digital platforms.
- **Collaborative Networks and Repositories**: These would house UX research, case studies, and resources that can be accessed by both academics and practitioners, ensuring that knowledge flows freely across the field.
- **Conferences, Workshops, and Forums**: Regular events that bring together academia and industry allow for dynamic exchanges of ideas, fostering collaboration and innovation.

By building a culture of openness, the UX field can avoid duplicated efforts, stay on the cutting edge of innovation, and ensure that both researchers and practitioners are contributing to a shared body of knowledge.

In summary, it is hoped that the proposed framework provides a comprehensive roadmap for addressing the critical gaps between academia and practice in UX. Through collaboration, integrated education, ongoing professional development, feedback loops, and open knowledge sharing, it ensures that both academic research and practical application are aligned, leading to faster innovation and a more cohesive, effective UX profession. This framework not only bridges the gap but also establishes a sustainable model for continuous learning, adaptation, and growth in the UX field.

Future Trends: The Impact of AI in UX Practice

Nielsen (2024), in his "10 fundamental insights for UX," asserts that "AI will revolutionize how we conduct UX work by dramatically boosting our productivity, enhancing our deliverables, and expanding our capabilities through generative user interface creation." Nielsen identifies AI as the latest major advancement in UX, starting with the release of ChatGPT 4 in 2023.

Nielsen sees two dimensions where AI will transform UX: (1) the intent-based outcome specification, and (2) the empowerment of individualization in UX. While the intent-based outcome specification is a revolutionary interaction paradigm, the individualization in the UX started from designing for target audience, to creating personas, to personalization, now to AI creating "individualized user experiences through generative UI." Nielsen predicts, "in a few years, generative UI will likely be able to create interfaces specifically for individual users on the spot, tailored to what that person needs right now."

As AI continues to evolve, the following trends are anticipated to reshape the UX landscape:

1. **AI-Driven User Research Tools**: AI will automate user research, simulating user behavior based on data, enabling UX professionals to predict interactions and optimize design without extensive manual testing.
2. **Personalized User Experiences**: AI will enable real-time adaptation of interfaces, creating more intuitive and engaging digital products. This "generative UI" (Nielsen, 2024) will enhance user satisfaction by tailoring interactions to individual preferences and needs.
3. **AI-Assisted Design Processes**: AI will streamline the design process by generating and iterating on prototypes, optimizing designs based on user data, and improving both time efficiency and outcomes.

In UX research, AI's ability to analyze large datasets and uncover patterns will accelerate the research phase, allowing practitioners to focus on strategic tasks and enhance project efficiency.

However, integrating AI into UX presents challenges, including ensuring AI-generated solutions remain ethical and user-centered. As Nielsen (2024) reiterates, "The underlying principles of UX are rooted in human behavior and cognition, not machine functionality." He advocates for using AI to augment, rather than replace, human intellect, noting that "AI can generate a flood of ideas, but humans can be better at prioritizing them and figuring out how to combine them into a design strategy. So, synergy is really the way to use AI in UX."

Academia and UX practitioners are urged to follow professional ethics and recognize the ethical guidelines on the use of AI, including but not limited to, European Union's (2019) "Ethics guidelines for trustworthy AI" and UNESCO's (2021) "Recommendation on the ethics of artificial intelligence." The collaboration between academia and industry will be crucial in developing best practices for leveraging AI in ways that benefit both users and businesses.

Summary

This chapter discusses the critical steps needed to bridge the gap between academic research and applied practice in UX. It proposes a structured framework that emphasizes collaboration, integrated education, continuous professional development, and knowledge sharing. At the end, it highlights the role of generative AI in shaping the future of UX, with AI-driven tools expected to revolutionize user research, personalized experiences, and design processes. Ultimately, this chapter underscores the importance of fostering strong connections between theory and practice, ensuring that UX remains an innovative and impactful discipline that evolves alongside emerging technologies.

References

European Commission. (2019). *Ethics guidelines for trustworthy AI. High-level expert group on artificial intelligence.* Retrieved from https://digital-strategy.ec.europa.eu/en/library/ethics-guidel ines-trustworthy-ai

Nielsen, J. (2024, August 15). 10 foundational insights for UX. https://jakobnielsenphd.substack. com/p/10-ux-insights

Thrift, J., & Tang, R. (2018). Teaching user experience (UX) in LIS programs and iSchools in North America: Challenges and innovations. In *Proceedings of ALISE 2018 Conference*, pp. 169–174.

UNESCO. (2021). *Recommendation on the ethics of artificial intelligence.* Retrieved from https:// www.unesco.org/en/artificial-intelligence/recommendation-ethics

Yu, F., Ruel, L., Tyler, R., Xu, Q., Cui, H., Karanasios, S., & Mostafa, J. (2020). Innovative UX methods for information access based on interdisciplinary approaches: Practical lessons from academia and industry. *Data and Information Management, 4*(1), 74–80.